Energy

D0317372

10
minutes

Test your knowledge

1 Types of energy:

- ***a*** A mug of hot coffee has .... energy.
- ***b*** A radio gives out .... energy.
- ***c*** A moving bus has .... energy.

2 An electric kettle transfers .... energy to .... energy. Since we can hear it making a noise as the water starts to boil, we know that some energy is wasted as .... energy.

3 An ordinary light bulb uses 100 J of electrical energy per second, and produces 5 J of light energy each second. A 'low energy' bulb uses 20 J of electrical energy per second, and produces 5 J of light energy each second. Which bulb is the most efficient?

4 A boy lifts a pile of books off the floor onto a shelf 2 metres high. The force needed to lift them is 60 N.

- ***a*** Calculate the work done by the boy.
- ***b*** How much energy does the boy transfer to the books?

5 Two boys of the same weight ride identical bikes up the same hill. The journey takes Tim 4 minutes, but Jack gets there in $3\frac{1}{2}$ minutes. Which boy produces the most power in his climb?

6

- ***a*** Coal, oil and nuclear fuels are examples of energy resources which are limited and will one day run out. They are known as .... energy resources.
- ***b*** In a coal-fired power station, coal is burnt to heat water. The steam produced turns the .... , which in turn drive the generator, which produces ....

Answers

1 thermal (or heat) / sound / kinetic **2** electrical / thermal (or heat) / sound **3** low energy bulb **4** a) 120 J b) 120 J **5** Jack **6** a) non-renewable b) turbines / electricity

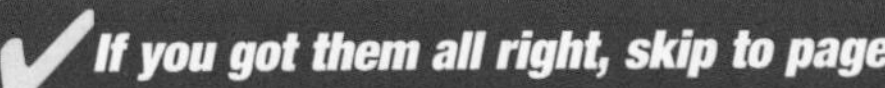
If you got them all right, skip to page 4

Energy

Improve your knowledge

1 Energy comes in many different forms. The main ones are:

- thermal or heat energy (e.g. from the sun)
- light energy (e.g. from a light bulb)
- sound energy (e.g. from a radio)
- chemical energy (e.g. in a battery)
- elastic or strain energy (e.g. in a stretched spring)
- electrical energy (e.g. used by a TV set)
- gravitational potential energy (due to height – e.g. bike at top of hill)
- kinetic or movement energy (e.g. a car)
- nuclear energy (e.g. used in a power station).

Energy is measured in joules (J).

2 Energy cannot be created or destroyed, but it can be **transferred** from one form to another, e.g. a television set turns electrical energy into light and sound energy.

3 However, energy transfers are not perfect, and some energy is wasted as **non-useful energy**, e.g. in a television set a little energy is turned into heat, and the back of the set feels warm. So we don't have as much useful energy produced by the television (as light and sound) as we put into it (as electrical energy). The less energy is wasted, the more efficient a device is.

The **efficiency** of a device is the fraction of the energy put in which is transferred to useful work.

4 Imagine we push a book a distance d along a table, and the force we need to push it with is F. The energy we've had to use (or the energy transferred from us to the book) is called the work done. It is measured in joules (J).

work done = force (F) x distance moved (d)

work done = energy transferred

Physics

Cathy Walters,
Abbey Tutorial College
Series Editor, Kevin Byrne

Where to find the information you need

SUCCESS OR YOUR MONEY BACK

Letts' market leading series GCSE in a Week gives you everything you need for exam success. We're so confident that they're the best revision books you can buy that if you don't make the grade we will give you your money back!

HERE'S HOW IT WORKS

Register the Letts GCSE in a Week guide you buy by writing to us within 28 days of purchase with the following information:

- Name
- Address
- Postcode
- Subject of GCSE in a Week book bought
- Probable tier you will enter

Please include your till receipt

To make a **claim**, compare your results to the grades below. If any of your grades qualify for a refund, make a claim by writing to us within 28 days of getting your results, enclosing a copy of your original exam slip. If you do not register, you won't be able to make a claim after you receive your results.

CLAIM IF...

You're a Higher Tier student and get a D grade or below.
You're an Intermediate Tier student and get an E grade or below.
You're a Foundation Tier student and get an F grade or below.
You're a Scottish Standard grade student taking Credit and General level exams, and get a grade 4 or below.
This offer is not open to Scottish Standard Grade students sitting Foundation level exams.

Registration and claim address:
Letts Success or Your Money Back Offer, Letts Educational, Aldine Place, London W12 8AW

TERMS AND CONDITIONS

1. Applies to the Letts GCSE in a Week series only
2. Registration of purchases must be received by Letts Educational within 28 days of the purchase date
3. Registration must be accompanied by a valid till receipt
4. All money back claims must be received by Letts Educational within 28 days of receiving exam results
5. All claims must be accompanied by a letter stating the claim and a copy of the relevant exam results slip
6. Claims will be invalid if they do not match with the original registered subjects
7. Letts Educational reserves the right to seek confirmation of the Tier of entry of the claimant
8. Responsibility cannot be accepted for lost, delayed or damaged applications, or applications received outside of the stated registration / claim timescales
9. Proof of posting will not be accepted as proof of delivery
10. Offer only available to GCSE students studying within the UK
11. SUCCESS OR YOUR MONEY BACK is promoted by Letts Educational, Aldine Place, London W12 8AW
12. Registration indicates a complete acceptance of these rules
13. Illegible entries will be disqualified
14. In all matters, the decision of Letts Educational will be final and no correspondence will be entered into

Letts Educational
Aldine Place
London W12 8AW
Tel: 020 8740 2266
Fax: 020 8743 8451
e-mail: mail@lettsed.co.uk
website: http://www.letts-education.com

First published 1998
Reprinted 1998, 1999 (twice)
New edition 2000

British Library Cataloguing in Publication Data
A CIP record for this book is available from the British Library.

ISBN 1 84085 3433

Design, artwork and production by Gregor Arthur
at Starfish Design for Print, London
Editorial by Tanya Solomons

Printed in Italy

Letts Educational is the trading name of Letts Educational Ltd, a division of Granada Learning Ltd. Part of the Granada Media Group.

5 We could, of course, move the book quickly or slowly. **Power** is the rate of doing work – the quicker we do it, the greater the power (in watts).

$$\text{power} = \frac{\text{work done}}{\text{time taken}}$$

you must be able to use this equation

6 Electricity is the most convenient form of energy for us to use. It is generated in power stations.

In a traditional power station a **fossil fuel** (coal, oil, or gas) is burnt to heat water. The steam produced is used to turn turbines. The turbines drive the generators which produce electricity.

Fossil fuels are known as **non-renewable** energy resources, because once they are used up they cannot be replaced. Uranium, used as a fuel in nuclear power stations, is also a non-renewable energy resource.

Renewable energy resources will not run out. Examples are solar power, wind power, tidal power, hydroelectric power (from fast-flowing water), geothermal power (using the internal heat of the Earth) and wood (trees grow quickly enough to replenish supplies).

When deciding on a source of energy, factors to consider include:

- Pollution caused by burning fossil fuels increases both global warming and acid rain.
- Nuclear power stations are costly to 'decommission' safely at the end of their lives.
- Although the sun and wind are 'free', the large-scale equipment needed to generate a significant amount of electricity is expensive, and covers a large area of land.
- The sun and wind are not always available when needed!

Checklist

Are you sure you understand these key terms?

energy transfer / non-useful energy / efficiency / work done / power / fossil fuel / renewable energy resource / non-renewable energy resource

Now learn how to use your knowledge

1 ***a*** Josh, a keen skier, is sitting in a ski lift which is moving at a constant speed up a mountain. What type of energy is he gaining as he moves higher up the slope? **Hint 1**

....

b Describe the energy changes that take place when he leaves the top of the mountain and starts to ski down. Try to include both useful and non-useful types of energy. **Hint 2**

....

....

....

....

While eating lunch in the ski café, Josh looks at some information about the 2 ski lifts on that slope:

> Height of summit = 300 m
> (above base of lifts)
>
> Time to reach summit on blue lift = 3 minutes
>
> Time to reach summit on red lift = $3^1/_2$ minutes

c Josh knows that the force needed to lift him is 750 N (his weight). What is the work done in taking Josh from the base of the lifts to the top of the mountain? **Hint 3**

....

....

....

d

$$\text{power} = \frac{\text{work done}}{\text{time taken}}$$

What is the power required to lift Josh up the mountain in 3 minutes? **Hint 4**

....

....

....

e The red and blue lifts are identical, except for their colour, and the times they take to reach the summit. Which lift is more powerful? **Hint 5**

....

2 **a** Explain the difference between a renewable and a non-renewable energy resource.

....

....

....

b Coal is one of the most common energy resources used in the UK. Give one advantage and one disadvantage of starting to use wind turbines to generate electricity, instead of coal.

Advantage: ....

....

Disadvantage: ....

....

✔ Hints and answers follow

Energy

Hints

1 As he's moving at constant speed, he isn't gaining kinetic energy.

What type of energy does an object have due to its height?

2 What type of energy does he have when at the top?

What 2 types of energy does he have while skiing halfway down?

What type of energy does he have when he reaches the bottom (at high speed!)?

What other types of energy may be produced, which may slow him down a bit?

3 The distance moved is just the height of the mountain, as the force is needed to lift him up, not to move him across.

You must learn the equation for work done.

4 The time must be in seconds to give an answer in watts.

3 minutes is 3 x 60 seconds.

5 Remember, the faster you do the work, the more powerful you are!

Answers

1 a) gravitational potential energy b) he loses gravitational potential energy, which is transferred to kinetic energy (useful) and a little heat and sound due to friction (non-useful) c) 225 000 J d) 1250 W e) blue **2** a) A renewable energy resource will not run out. A non-renewable resource is one which there is a limited amount which cannot be replaced b) *Advantage:* Wind is a renewable resource and will not run out. (*or* Using wind energy does not cause acid rain or global warming.) *Disadvantage:* The wind is unreliable. (*or* 'Farms' of wind turbines are very large and spoil the countryside.)

Test your knowledge

1. If a spoon at room temperature (20°C) is put into a mug of hot coffee at 70°C, some heat energy will move from the .... to the ....

2. Heat travels through the water in an electric kettle by convection. Name 2 other methods of heat transfer.

3. Which method of heat transfer is most effective in metals?

4. Which method of heat transfer works well in liquids and gases, but not in solids?

5. One house is painted black, another is painted white. Which house will absorb more heat from the sun?

6. A duvet is good at keeping the heat in a bed because it **traps air**. This reduces heat loss by 2 methods – which ones?

Answers

1 coffee / spoon **2** conduction / radiation
3 conduction **4** convection **5** black house
6 conduction and convection

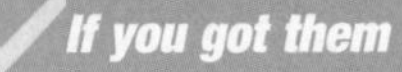

If you got them all right, skip to page 10

Heat transfer

1. If 2 objects are at different temperatures, some heat energy will be transferred from the one at the higher temperature to the one at the lower temperature, *if* there is a way for this to occur.

   If different parts of a substance are at different temperatures, heat energy will be transferred from the parts at the higher temperature to the parts at the lower temperature.

2. These transfers of heat occur mainly by **conduction**, **convection** and **radiation**.

3. **Conduction** involves the transfer of heat energy within a substance, without the substance itself moving. As the particles in the substance get hotter they vibrate more. They bump into the particles around them, passing on the heat.

   Metals are the best conductors of heat. Other solids and liquids are fairly poor conductors (often called insulators). Gases are extremely poor conductors.

4. **Convection** can only occur in liquids or gases, because these substances can flow.

   e.g.

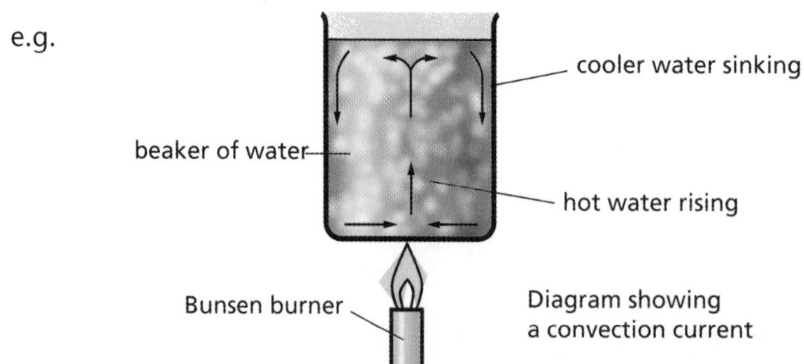

The hot parts of the water have a lower density and so will rise, carrying the heat with them. This allows cooler water to move in and be heated.

5 **Radiation** is energy emitted from an object in the form of a wave (usually infrared radiation). All objects emit radiation, but the hotter they are the more they emit. Radiation will travel through gases or a vacuum (empty space). When it reaches another object, some of the radiation will be absorbed by the object.

Dull (matt) or black objects are better at radiating heat than shiny or white ones.

Dull (matt) or black objects are also better at absorbing radiant heat (from the sun or other sources of heat) than shiny or white ones.

6 **Insulation** can be used to reduce heat transfer. An insulator is a poor conductor. Often insulating materials (e.g. the lagging on a hot water tank) trap air, which is a very poor conductor. Air is only a useful insulator if it is trapped, because if it can move around, then heat can be transferred by convection.

For example, to reduce heat loss from a house, the roof, windows and walls can be insulated. Loft insulation is made of a material designed to trap air. Double glazing traps air (or in some cases creates a partial vacuum) between 2 layers of glass, and cavity wall insulation traps air in a foam between 2 layers of bricks. Insulating the loft is the most important of these measures, as warm air inside the house is carried towards the roof by convection currents.

Checklist

Are you sure you understand these key terms?

conduction / convection / radiation / insulation

Now learn how to use your knowledge

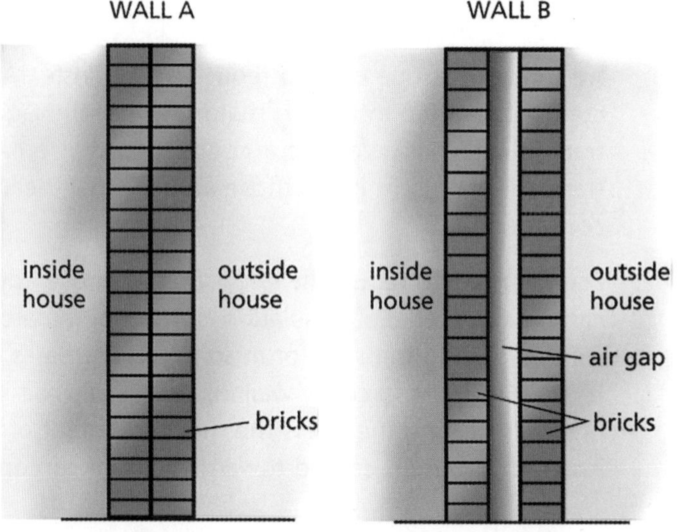

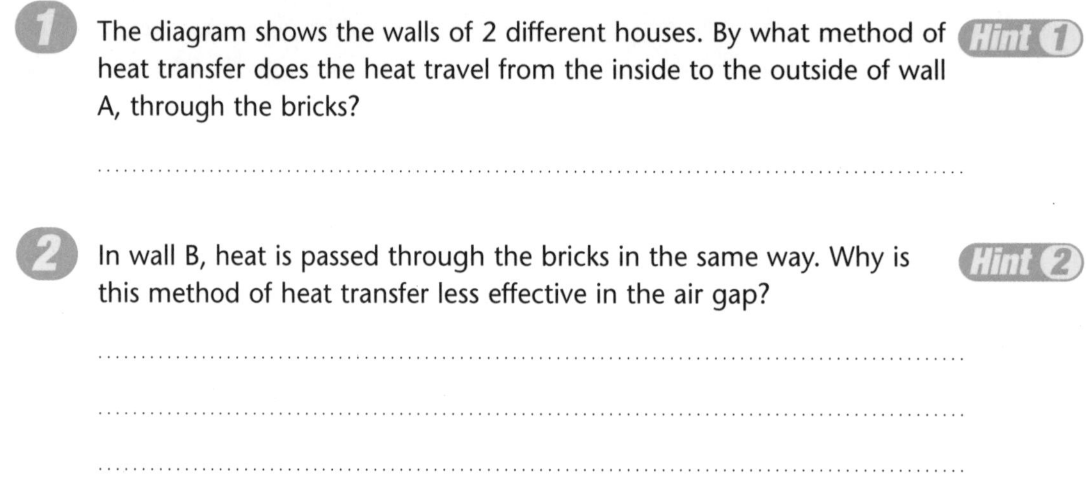

1. The diagram shows the walls of 2 different houses. By what method of heat transfer does the heat travel from the inside to the outside of wall A, through the bricks? Hint 1

....

2. In wall B, heat is passed through the bricks in the same way. Why is this method of heat transfer less effective in the air gap? Hint 2

....

....

....

3 It is possible to inject an insulating foam into the gap in wall B. How would this reduce heat loss through the wall? Hint 3

....

....

....

4 In general, more heat is lost through the roof of a house than through the walls. Why? Hint 4

....

....

....

5 Explain how this heat loss can be reduced. Hint 5

....

....

....

....

6 A scientist has decided to paint his own (currently white) house black as an experiment in heat loss. Would you expect his house to lose more or less heat now at night – and why? Hint 6

....

....

....

Hints and answers follow

Heat transfer

Hints

1 Brick is a solid.

Only one of the methods of heat transfer will work in solids.

2 Brick is a solid, but air is a gas.

3 The problem with wall B is that convection currents can move in the air gap. How does the foam stop this?

4 The roof is at the top of the house. Which kind of heat transfer tends to move heat upwards?

5 What kind of insulation is used to reduce heat loss from the house to the loft?

What is trapped in this insulation?

How does this reduce the heat transfer?

6 Which emits most radiant heat – black or white paint?

Answers

1 conduction **2** air is a gas, so a worse conductor than a solid **3** foam traps the air in tiny pockets, preventing convection currents in the air gap **4** hot air tends to rise towards the top of a house by convection **5** loft insulation traps air, making a barrier through which heat cannot easily pass by conduction or convection **6** the black house loses more heat, as black surfaces emit more radiant heat than white ones

Light and the electromagnetic spectrum

Test your knowledge

1 A ray of light hits a mirror at angle **a** and is reflected at angle **b**:

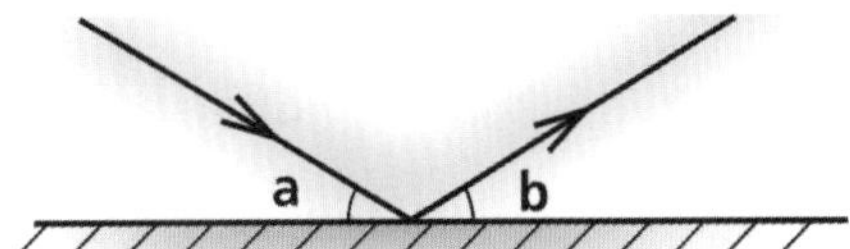

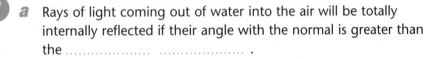

Is angle **a** bigger, smaller, or the same size as angle **b**?

2 A ray of light will change direction when it enters a glass block from the air, unless it hits the block at ....

3 *a* Rays of light coming out of water into the air will be totally internally reflected if their angle with the normal is greater than the ....

b State one practical use of total internal reflection.

4 In the electromagnetic spectrum, which kind of wave

a has the longest wavelength?

b has the highest frequency?

c is used in grills, in radiant heaters, and to operate videos by remote control?

Answers

1 the same size **2** 90° (right angles) **3** a) critical angle b) fibre optics or binoculars or periscopes **4** a) radio waves b) gamma waves c) infrared

✓ If you got them all right, skip to page 16

Light and the electromagnetic spectrum

Improve your knowledge

1 When light is **reflected** from a plane (flat) mirror, the angle at which it is reflected is the same size as the angle at which it hits the mirror:

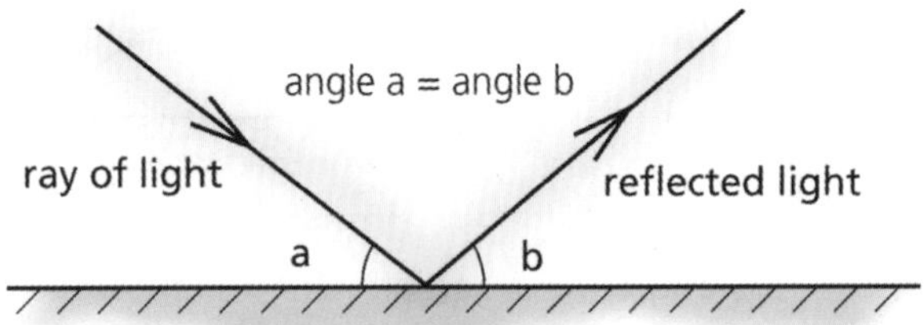

2 Light is **refracted** when it passes from one transparent substance to another. The speed of the light is different in different substances, and this causes the light to change direction (unless it hits the boundary at right angles).

Rays of light going from air into glass, water, or perspex are slowed down and bent towards the normal.

Rays of light going out of glass, perspex or water and into air will speed up and bend away from the normal:

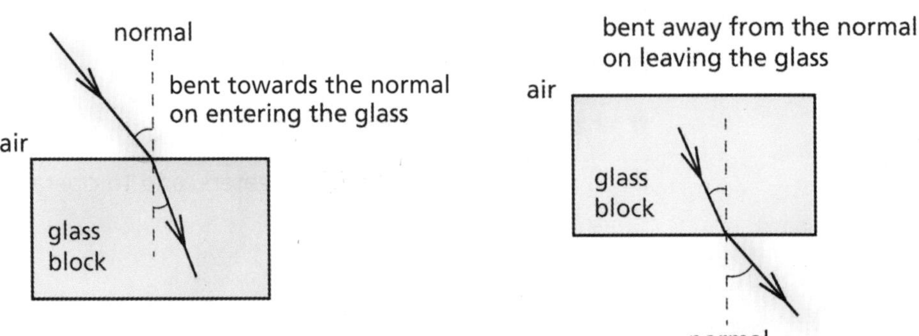

*the **normal** is always drawn at right angles to the surface, to help us measure the angles*

3 Rays of light coming out of glass, perspex or water into the air will be **totally internally reflected** if their angle with the normal is greater than the **critical angle**:

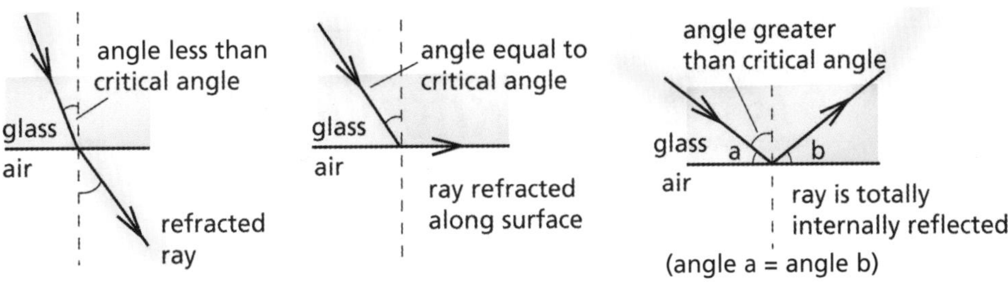

A ray of light can be transmitted by a thin optical fibre, because if it hits the side it is reflected by total internal reflection.

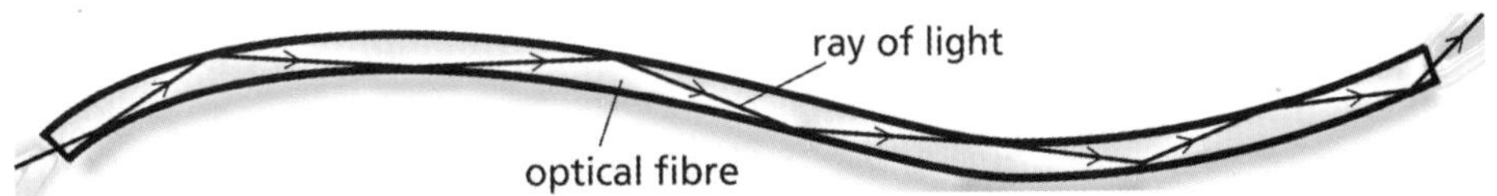

4 The **electromagnetic spectrum** consists of waves which are like light, but have different wavelengths and so have different properties.

Type of wave	Wavelength	Frequency	Uses
gamma rays	shortest	highest	to sterilise surgical instruments
X-rays	↓	↓	to 'photograph' fractures
ultraviolet light			sun lamps
visible light			for humans to see
infrared light			in heaters and toasters
microwaves			cooking in microwave ovens
radio waves	longest	lowest	transmitting radio and TV signals

Dangers: Large doses of ultraviolet, X-rays or gamma rays can kill cells, while smaller doses may cause cancer. Microwaves may kill cells by overheating, and too much infrared (felt as heat) can cause burns.

Checklist

Are you sure you understand these key terms?

reflected / refracted / normal / total internal reflection / critical angle / electromagnetic spectrum

Now learn how to use your knowledge

Use your knowledge

1 *a* Place the following in order, starting with the shortest wavelength:

Hint 1

ultraviolet light, visible light, microwaves, X-rays

....

b Give one use and one danger of gamma rays.

Use: ....

Danger: ....

2 A ray of light enters a perspex block at an angle as shown below:

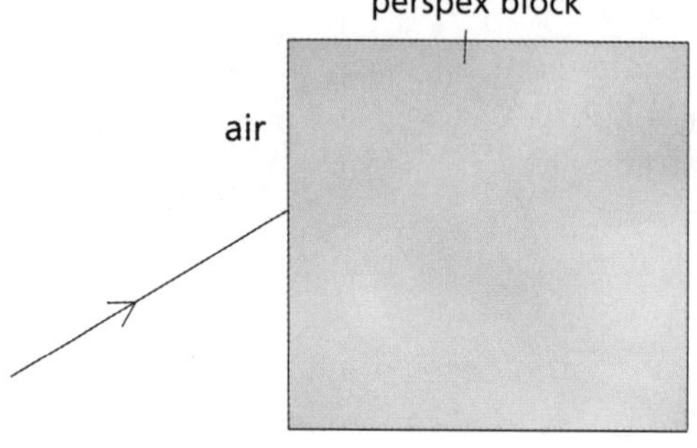

a On the diagram, sketch the path of the ray inside the block. Hint 2

b What is the name given for what happens as the ray enters the block?

....

A ray of light passes through a glass prism as shown below. The critical angle for glass is 42°.

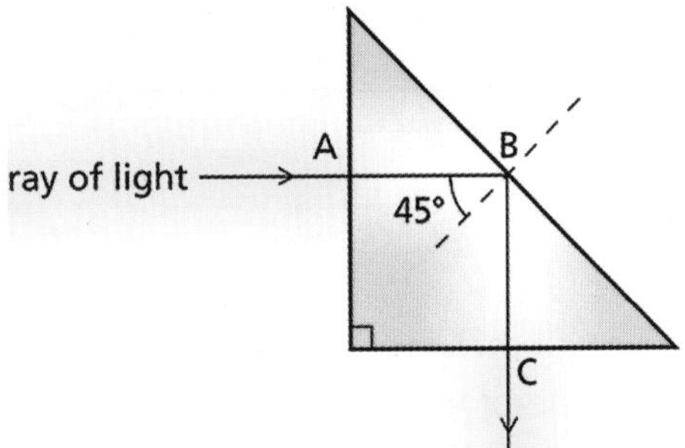

c Why doesn't the ray of light change direction when it enters the prism? Hint 3

....

d State what happens to the ray at B.

....

e Explain why the light does not leave the prism at B. Hint 4

....

....

....

f In binoculars, two prisms are used to reflect light back in the direction it came from. Draw a diagram to show how this could be achieved. Hint 5

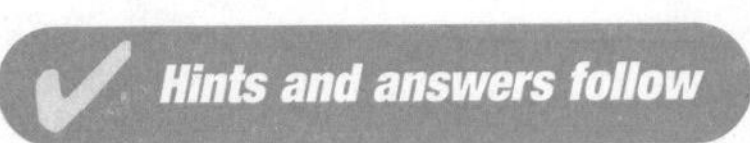

Light and the electromagnetic spectrum

Hints

1 You should learn all the parts of the electromagnetic spectrum, in order, and at least one use of each part – plus the main dangers.

2 First draw in the normal as a dotted line. Remember that rays *entering* perspex or glass are bent *towards* the normal.

3 At what angle does the ray hit the block at A?

4 What is the angle between the ray and the normal at B? The critical angle for glass is given earlier in the question.

5 Draw a prism like the one in the question. Add another prism so that the ray coming out of the first prism will go straight into it.

Make sure you have the second prism the right way round so the ray will be reflected back the way it came – towards the left side of the page. Finally draw in the path of the ray of light through the prisms.

Answers

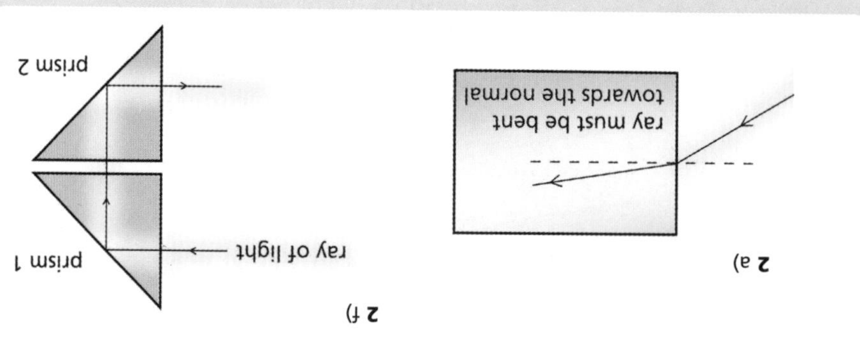

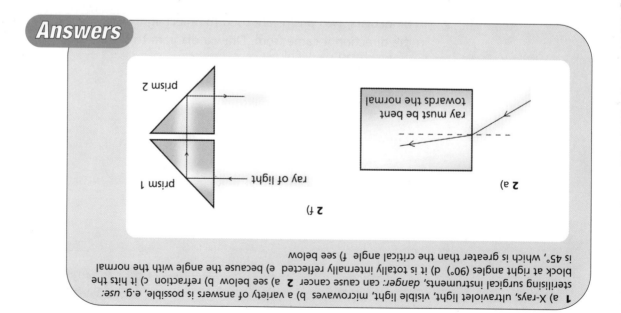

1 a) X-rays, ultraviolet light, visible light, microwaves b) a variety of answers is possible, e.g. *use:* sterilising surgical instruments, *danger:* can cause cancer **2** a) see below b) refraction c) it hits the block at right angles (90°) d) it is totally internally reflected e) because the angle with the normal is 45°, which is greater than the critical angle f) see below

Waves and sound

Test your knowledge

1. Give one example of a transverse wave.

2. Give one example of a longitudinal wave.

3. The distance between 2 adjacent peaks on a water wave is called the .... of the wave.

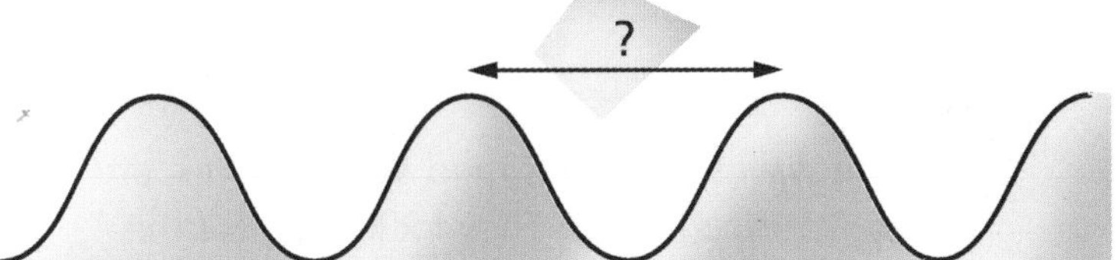

4. When water waves pass suddenly into shallower water they are slowed down. As a result of this they change direction. This is an example of ....

5. A .... wave travels through the air as a vibration of the air molecules.

6. **a** If the amplitude of a sound wave increases, it will sound ....

   b If the frequency of a sound wave increases, it will sound ....

7. What causes an echo?

8. Very high frequency sound waves, used in the prenatal scanning of babies, are known as ....

Answers

1 light waves / water waves / waves in a rope
2 sound waves / some waves in springs
3 wavelength **4** refraction **5** sound
6 a louder b) higher in pitch **7** the reflection of a sound wave **8** ultrasound

✔ If you got them all right, skip to page 22

Waves and sound

Improve your knowledge

1 Waves made by shaking a rope, ripples in water, and light waves are called **transverse** waves. They look a bit like this:

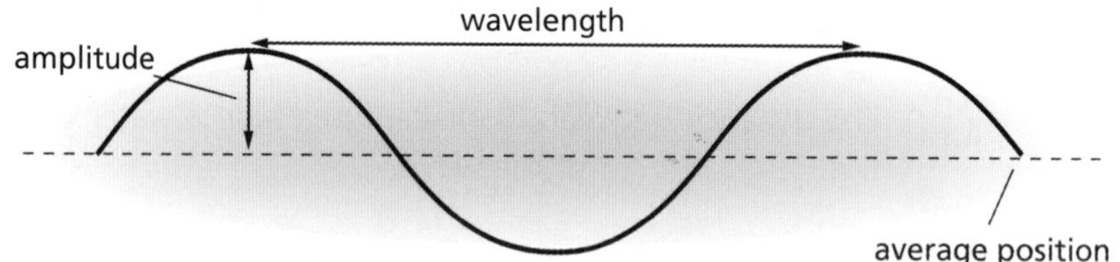

2 Some waves in springs, and sound waves, are a different kind of wave, called a **longitudinal** wave. They look a bit like this:

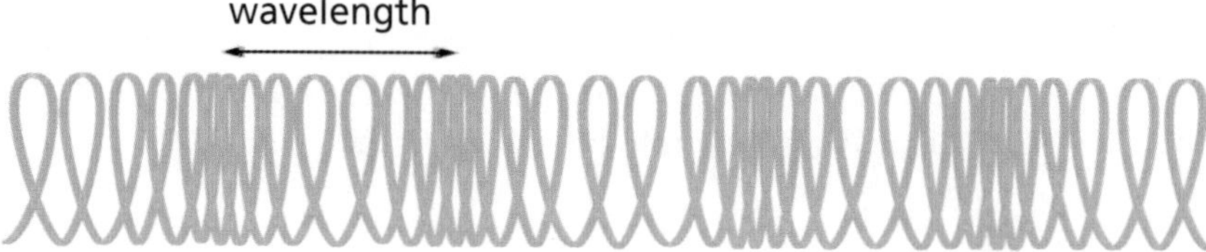

3 The **amplitude** of a wave is the maximum displacement from the average position.

The **wavelength** of a wave is the distance between two peaks of the wave.

The **frequency** of a wave is the number of cycles (up and down) it completes every second. It is measured in hertz (Hz).

4 All waves have the following properties:

a They can be reflected.

b They can be refracted.

c They transfer energy without transferring matter, e.g. a wave pattern travels along a rope, but the rope itself does not move along.

5 A sound is produced by a vibrating object, e.g. a violin string, or the voice box in our throat. It travels through the air as a wave, vibrating the air molecules.

6 The bigger the *amplitude* of a sound wave, the louder the sound.

The higher the *frequency* of the sound wave, the higher the pitch of the note.

7 A sound wave can be reflected from a hard, flat surface, like a large wall or the back of a cave. When a sound wave is reflected back to us, we hear an echo.

8 **Ultrasound** is a sound wave with a very high frequency, too high for us to hear. It is reflected by body tissues and can be used to scan a pregnant woman to check on her unborn baby.

Pulses of ultrasound are also used for depth measurements in deep water. By measuring the time taken for the pulse to reach the bottom and be reflected back, the depth can be calculated.

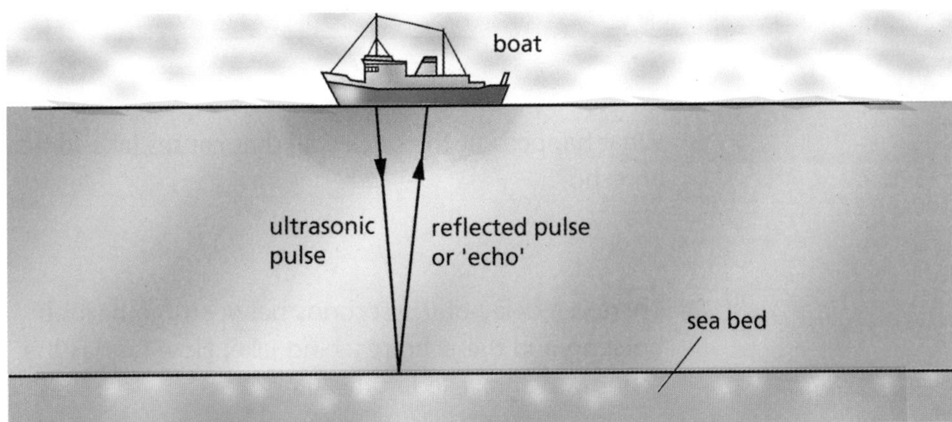

Checklist

Are you sure you understand these key terms?

transverse / longitudinal / amplitude / wavelength / frequency / ultrasound

✓ Now learn how to use your knowledge

Waves and sound

1 Jake bursts a balloon a little distance away from a brick wall. The noise it makes travels through the air as a sound wave. Jake hears an echo.

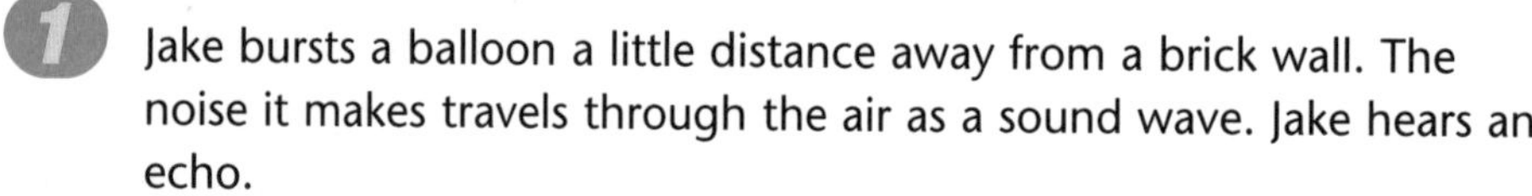

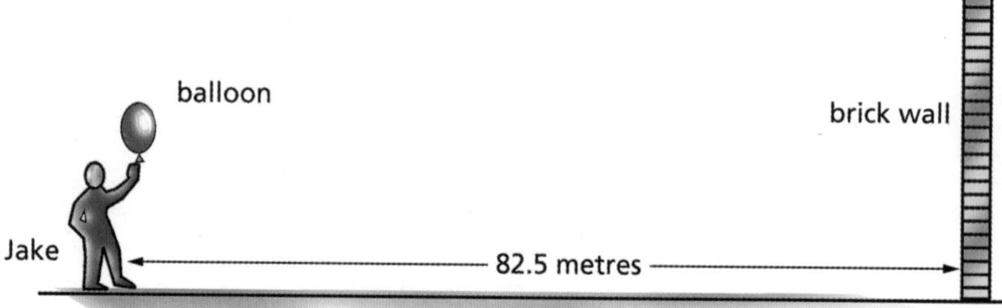

a What happens at the brick wall that causes Jake to hear an echo?

....

b There is a delay of 0.5 seconds between the balloon bursting and the echo reaching Jake. How far has the sound wave travelled in this time?

....

c Use the information given to calculate the speed of sound in air.

$$\text{speed} = \frac{\text{distance travelled}}{\text{time taken}}$$

....

....

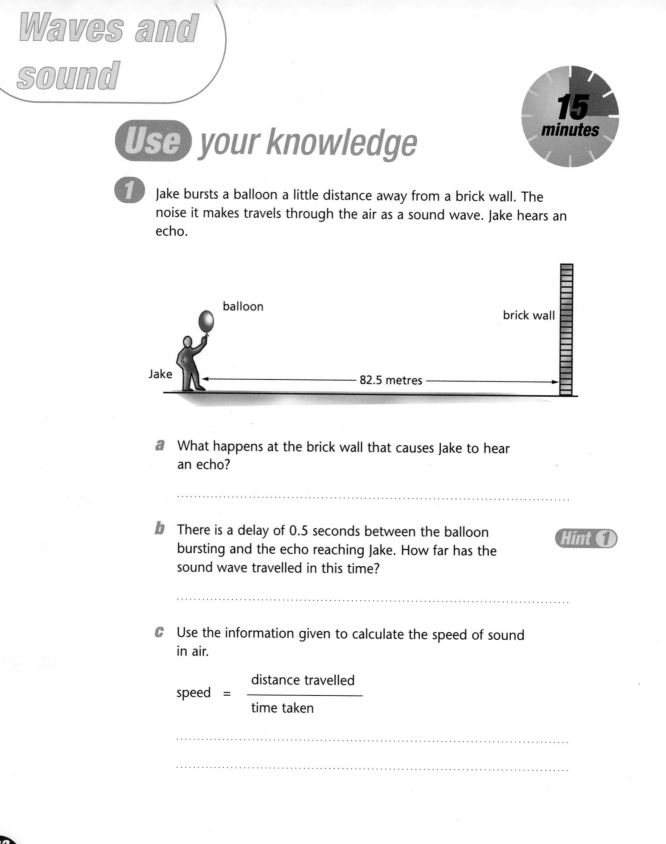

d Echoes can also occur with ultrasound. Give one use of ultrasonic echoes.

....

....

2 Some water waves are moving across a tank of water. The tank, viewed from the side, is shown below. The waves are moving to the right.

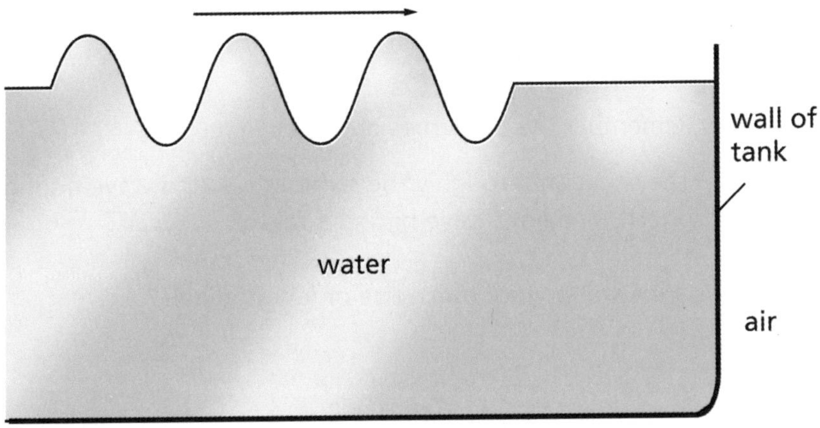

wall of tank

water

air

a Mark on the diagram

1) the wavelength of the waves

2) the amplitude of the waves.

Hint 2

b What will happen to the waves when they hit the wall of the tank?

Hint 3

....

....

c Is this wave more similar in nature to a light wave or a sound wave? Give a reason for your answer.

Hint 4

....

....

Hints and answers follow

Waves and sound

Hints

1 Remember that the sound wave has travelled to the wall, and back.

2 Remember that the amplitude is the greatest distance from the *average* position.

3 Remember the properties of waves.

The situation is roughly the same as a sound wave hitting a wall or cliff, or a light wave hitting a mirror.

4 Is the water wave transverse or longitudinal?

Answers

1 a) the sound wave is reflected b) 165 m c) 330 m/s d) finding the depth of water (or scanning unborn babies) **2** a) see below b) they will be reflected back to the left c) a light wave, because it is transverse

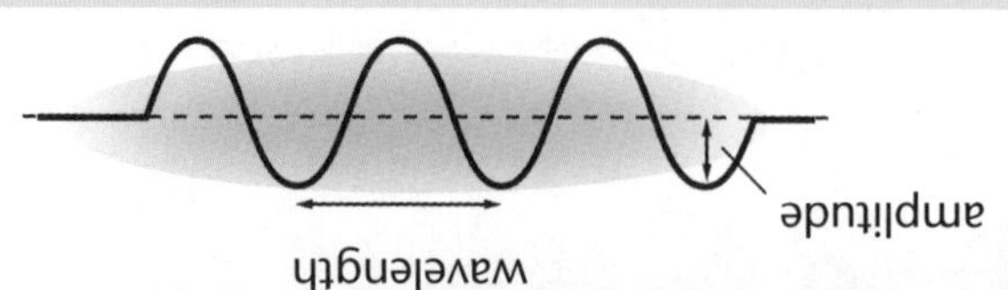

Earth and beyond

Test your knowledge

1 The Moon is constantly orbiting the ....

The Earth is constantly orbiting the ....

2 The Sun, Earth, and other planets are together known as the ....

3 The Sun is our nearest .... We can see the planets because they .... light from the Sun.

4 Our Sun is part of a group of millions of stars called the ....

5 The planets remain in orbit around the Sun because of the force of ....

This force depends on the mass of the planet, and also on its .... from the Sun.

6 Man-made objects which orbit the Earth to send television signals around the world or to help with weather observations are called ....

Answers

1 Earth / Sun **2** Solar System **3** star / reflect
4 Milky Way galaxy **5** gravity / distance
6 satellites

✔ *If you got them all right, skip to page 28*

Improve your knowledge

1 The Earth is constantly **orbiting** the Sun, following a path shaped like an ellipse (which is a slightly squashed circle). It takes a year for the Earth to go round the Sun once.

At the same time, the Moon is orbiting the Earth. It takes about 4 weeks to go round once.

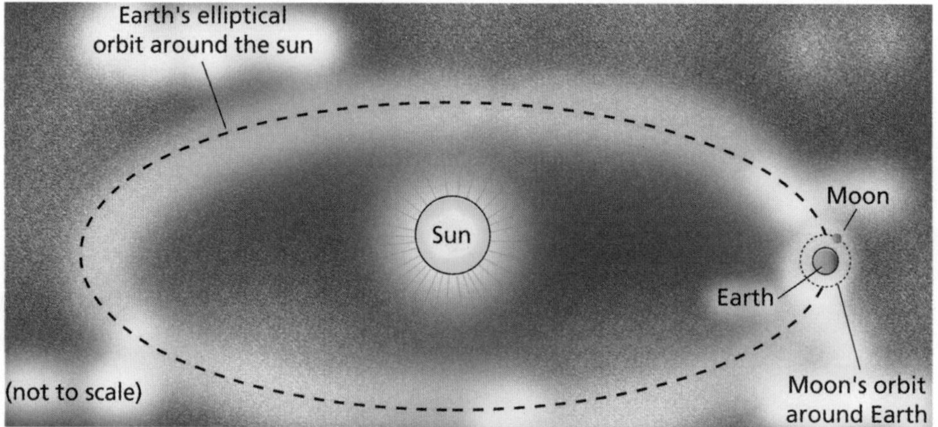

2 The **Solar System** consists of the Sun, and the Earth and other planets which orbit it. Some of the other planets also have moons. In general, the closer a planet is to the Sun, the hotter it will be. (See diagram on next page.)

3 The Sun is our nearest star. It may seem to us to be much bigger than the other stars, but that is just because it is hundreds of thousands of times closer than they are.

A **star** gives out light and heat. **Planets** and **moons** do not give out light, but we can see them if they reflect light from the Sun.

4 A **galaxy** is a vast group of stars. Our Sun is just one of millions of stars which make up the Milky Way galaxy.

The **universe** is made up of at least a billion galaxies.

5 The planets are held together in the Solar System, and the stars are held together in the galaxy, by the force of **gravity**. This force acts between all objects with a mass – but it is stronger if the objects have a large mass, or are close together. Without gravity the planets would not orbit the Sun; they would just keep moving in a straight line through the galaxy.

6 Artificial **satellites** are man-made objects which can be made to orbit the Earth. They have a variety of uses – including monitoring the weather and relaying TV signals from one part of the world to another.

Simplified diagram of the Solar System

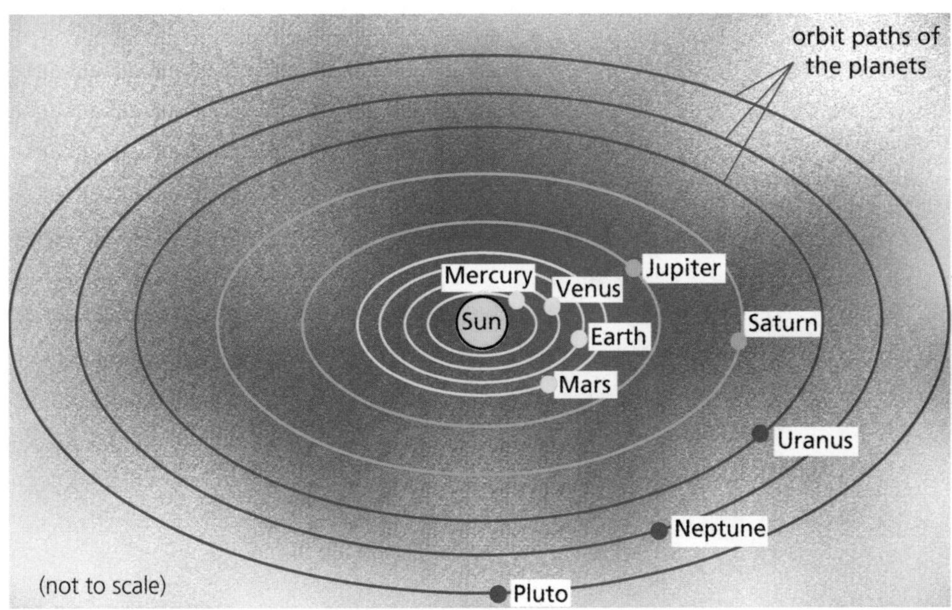

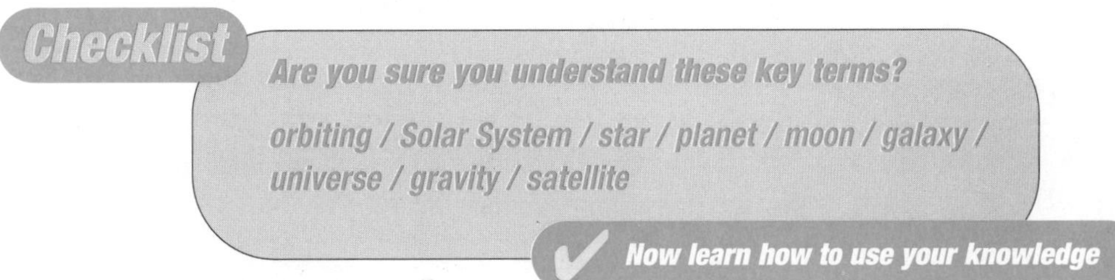

Checklist
Are you sure you understand these key terms?
orbiting / Solar System / star / planet / moon / galaxy / universe / gravity / satellite
Now learn how to use your knowledge

Earth and beyond

1 Use the following list to answer the questions below:

Earth, Moon, Milky Way galaxy, Sun, Solar System

Hint 1

Which of the above is

a the largest? ....

b the smallest? ....

c a star? ....

2

Planet name	**Distance from Sun** (millions of km)	**Mass of planet** (compared with Earth)
Jupiter	780	320
Neptune	4500	17

a Which of the two planets in the table would you expect to have the coldest surface temperature? Why?

Hint 2

....

....

....

b Both planets are held in their orbits around the Sun by the force of gravity. Give 2 reasons why the force holding Jupiter in orbit is greater than that holding Neptune in orbit.

Hint 3

....

....

....

....

3 SIPA 2 is in orbit around the Earth, collecting data used for weather forecasting.

PEL 1 is also in orbit around the Earth, relaying television signals from one part of the world to another.

a What is the name for objects such as SIPA 2 and PEL 1 in orbit around the Earth?

....

b PEL 1 is in a geostationary orbit. This means it orbits the Earth at the same rate as the Earth itself turns round, so it always stays above the same point on the Earth's surface.

PEL 1 sends television signals to satellite dishes on houses. Why is it important that PEL 1 always stays in the same position above the Earth's surface? **Hint 4**

....

....

....

c Why is SIPA 2 *not* placed in a geostationary orbit? **Hint 5**

....

....

....

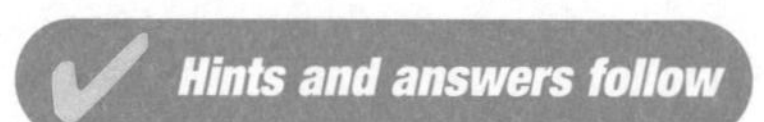

Earth and beyond

Hints

1 Read the *Improve your knowledge* section again if you are unsure. You must know what each of these is.

2 Which piece of information in the table will affect how warm the planet is?

Planets get their heat (mostly) from the Sun, so the closer they are to the Sun the warmer they will be.

3 The force of gravity between two objects is greater the larger their masses, and the closer together they are.

The closer together the Sun and the planet are, the greater the force of gravity between them.

The greater the mass of the planet, the stronger the force of gravity between it and the Sun.

4 A satellite dish must be pointing at the satellite to receive the signal.

What would happen if the satellite was moving relative to the Earth?

5 Would the weather information collected by SIPA 2 be more useful if collected from one area of the Earth, or from many areas?

Does a geostationary orbit allow this?

Answers

1 a) Milky Way galaxy b) Moon c) Sun **2** a) Neptune, because it is further from the Sun b) Jupiter has a greater mass, and is closer to the Sun **3** a) satellites b) because satellite dishes on houses always point at one fixed point in space, and cannot constantly move to track a satellite c) because more weather information can be collected from around the world if the satellite is not held in a fixed position

Motion

Test your knowledge

1 *a* A car travels 90 miles in 2 hours. Calculate its (average) speed in miles per hour.

b A man walks 900 m in 10 minutes. Calculate his speed in m/s.

2 Which is moving faster, marble A or marble B?

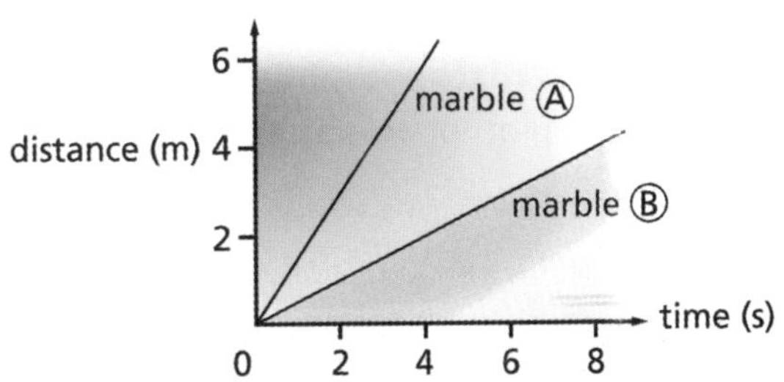

3 Pigeon A flies straight from Birmingham to Coventry at 15 m/s. Pigeon B flies straight from Coventry to Birmingham at 15 m/s. Do both birds have the same velocity? Why?

4 A car increases its velocity from 10 m/s to 30 m/s in 10 seconds. Calculate its acceleration.

5 Is the car speeding up, slowing down, or moving at a constant speed?

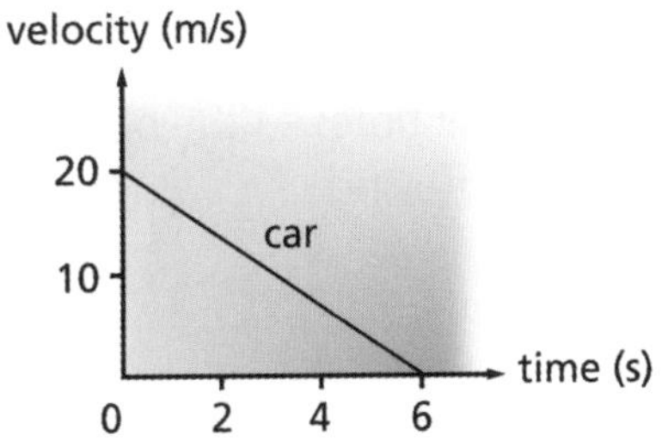

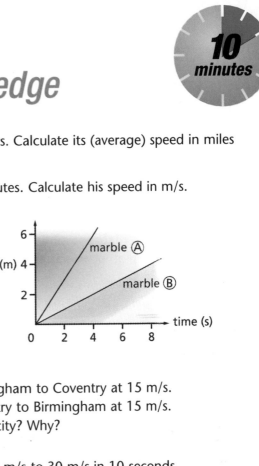

Answers

1 a) 45 mph b) 1.5 m/s (don't forget that 10 minutes is 10 x 60 seconds!) **2** A **3** no, as velocity depends on speed and direction **4** 2 m/s^2 **5** slowing down

If you got them all right, skip to page 34

Motion

1 **Speed** is usually measured in metres per second (m/s).

$$\text{speed} = \frac{\text{distance travelled}}{\text{time taken}}$$

Learn this

2 A **distance–time** graph is a 'plot' of an object's motion:

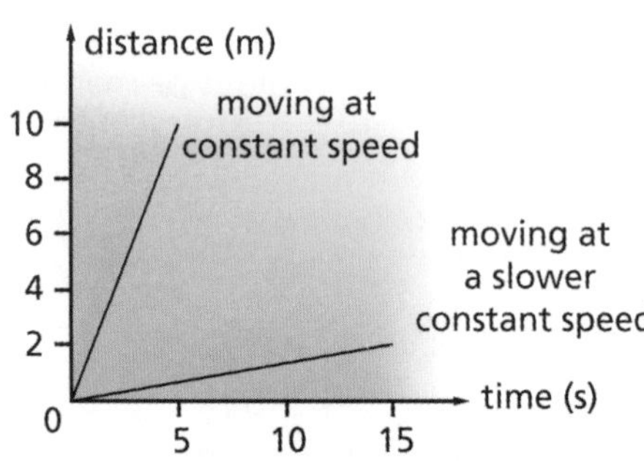

A straight line means the object is travelling at a constant speed.

The *steeper* the slope, the *greater* the speed.

You can calculate speed from a distance–time graph: it is the *gradient* of the graph.

If the graph is a horizontal line, the distance is the same at all times so the object is not moving.

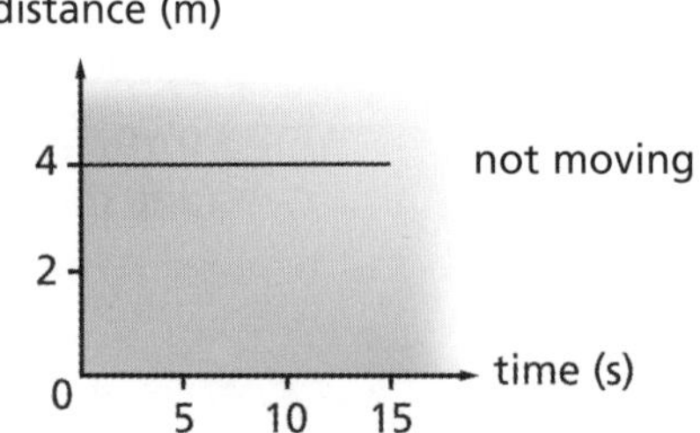

3 **Velocity** is the *speed in a certain direction*, in m/s like speed. So, when a car goes round a bend, its velocity will change even if its speed stays the same. Velocity may be described as 5 m/s *to the left*, for example.

4 If your velocity changes, you **accelerate**. The faster your velocity changes, the greater your acceleration.

$$\text{acceleration} = \frac{\text{velocity change}}{\text{time taken}}$$

Learn this

Acceleration is measured in m/s^2.

5 On a **velocity–time graph**, a *straight* line means the object has a constant acceleration. The steeper the slope, the greater the acceleration. A *horizontal* line means constant velocity.

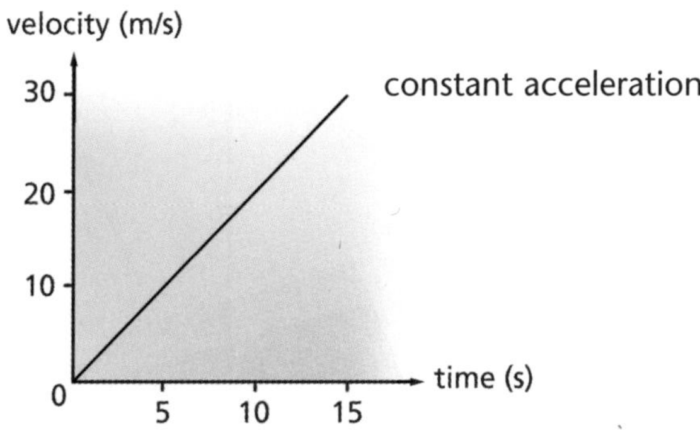

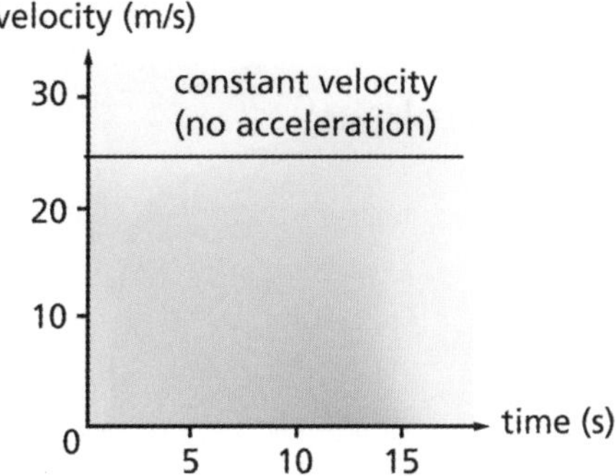

Are you sure you understand these key terms?

speed / velocity / acceleration

Now learn how to use your knowledge

Motion

The graphs show the movement of two people over a 20 second time period:

distance (m)
100
80
60
40
20
0
5 10 15 20
time (s)

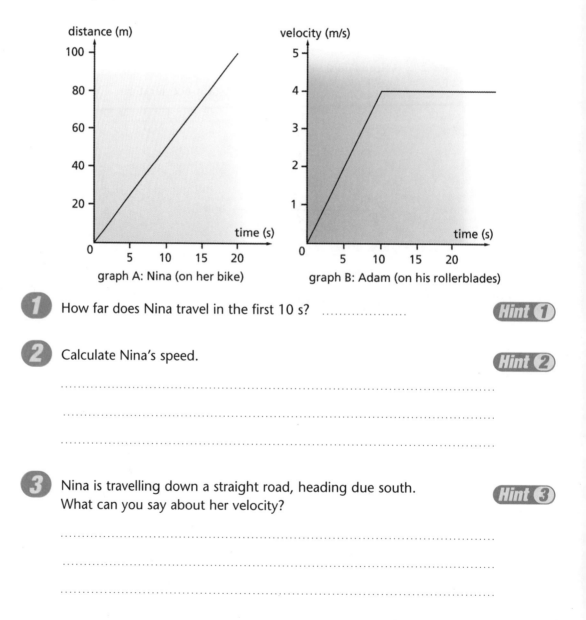

graph A: Nina (on her bike)

velocity (m/s)
5
4
3
2
1
0
5 10 15 20
time (s)

graph B: Adam (on his rollerblades)

1 How far does Nina travel in the first 10 s? .... **Hint 1**

2 Calculate Nina's speed. **Hint 2**

....

....

....

3 Nina is travelling down a straight road, heading due south. What can you say about her velocity? **Hint 3**

....

....

....

4 Sketch a **velocity–time** graph for Nina's motion. Hint 4

5 Now look at graph B. Describe Adam's motion Hint 5

a in the first 10 s.

....

b between 10 and 20 s.

....

6 What is Adam's velocity

a after 5 s? ....

b after 10 s? ....

c Calculate his acceleration between 5 and 10 seconds. Hint 6

....

....

....

Hints and answers follow

Motion

Hints

1 Read it off the graph!

2 Speed is the gradient of a distance–time graph.

Or use speed = distance / time.

3 Remember, velocity is speed in a certain direction.

If her speed and direction are both constant, her velocity will be constant.

4 What is her velocity?

Is it increasing, decreasing, or constant?

How do you show this on a graph?

5 a) What does a straight line on a velocity–time graph mean?

b) If the line is horizontal, this is a special case.

6 You should have **learnt** the equation for calculating acceleration. If not, look it up on page 33.

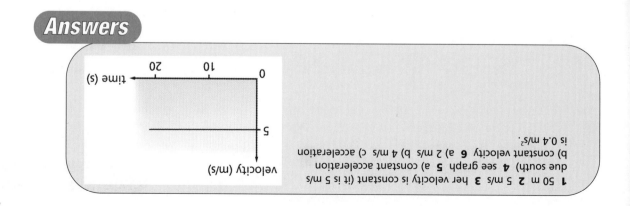

Answers

1 50 m **2** 5 m/s **3** her velocity is constant (it is 5 m/s due south) **4** see graph **5** a) constant acceleration b) constant velocity **6** a) 2 m/s b) 4 m/s c) acceleration is 0.4 m/s^2.

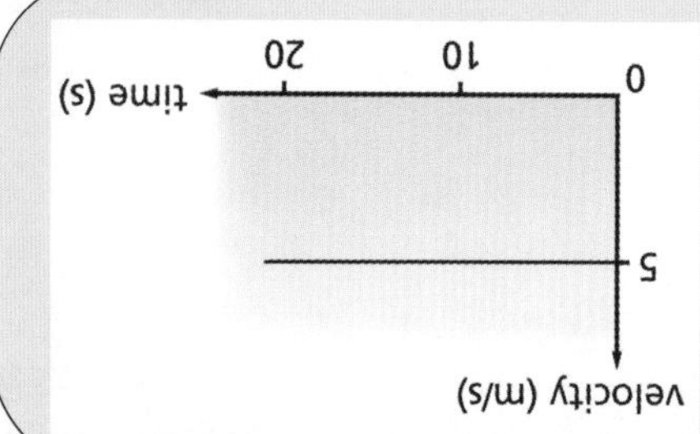

Test your knowledge

1 If balanced forces are acting on an object, the object will move at a .... speed.

2 3 trolleys are each pulled by a force of 3 N:

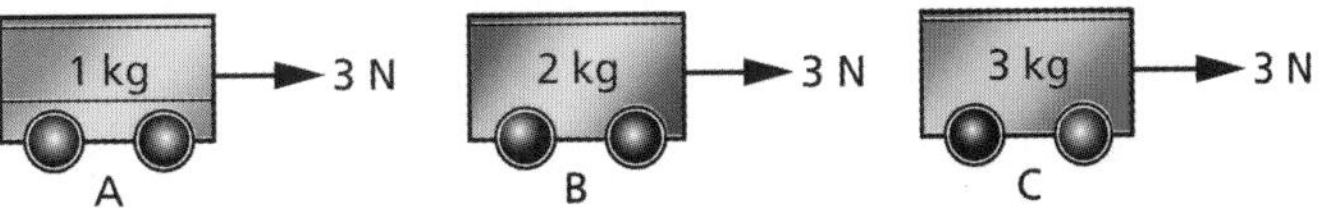

Which trolley will have the greatest acceleration?

3 A man is leaning on a wall. He puts a force of 10 N on the wall. The wall is also putting a force on the man – is it

a less than 10 N? ***b*** more than 10 N? ***c*** exactly 10 N?

4 A marble is rolling across the floor. If there were no forces on it, it would keep rolling at the same speed. In fact, it slows down and stops. Why?

5 Is the parachutist accelerating, decelerating, stationary, or moving with constant speed?

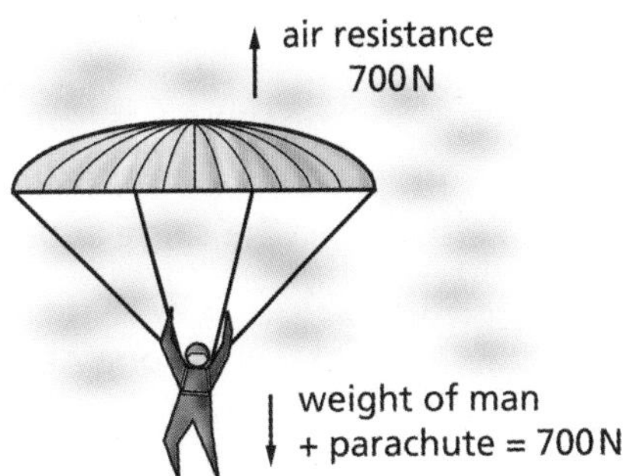

6 When calculating a car's stopping distance, you have to allow for the 'thinking distance'. What is this?

Answers

1 constant **2** A **3** exactly 10 N **4** friction **5** moving with constant speed **6** the time it takes the driver to react and depress the brake

If you got them all right, skip to page 40

Forces in action

Improve your knowledge

Balanced forces are forces which cancel each other out.

e.g. No overall force acts

If either no force, or balanced forces, act on an object, it won't change its speed. If it is stationary, it will remain stationary. If it was moving, it will continue moving at the same speed in the same direction.

e.g. A car moves at constant speed if the forward force due to the engine exactly balances the frictional forces.

If **unbalanced forces** (forces which do not cancel out) act on an object, the object will accelerate (speed up, slow down, or change direction).

e.g. The car will slow down. The **resultant force** is 200 N to the left.

The greater the resultant force, the greater the acceleration. But the greater the mass, the more force is needed to cause the same acceleration.

3 When one object (e.g. a mug) puts a force on another object (e.g. a table), the second object (table) always puts an equal force, in the opposite direction, on the first object (mug).

e.g. Force of mug on table = 1 N (its weight). Therefore, force of table on mug = 1 N (called the **reaction force**).

4 **Friction** is a force which acts to slow you down, when 2 solid surfaces slide across one another, or when an object is moving through water or air ('air resistance'). The direction of the frictional force is always the opposite direction to the motion.

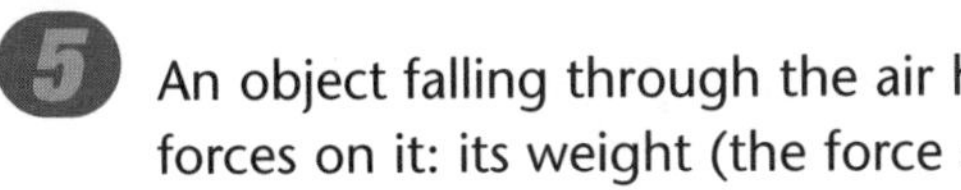

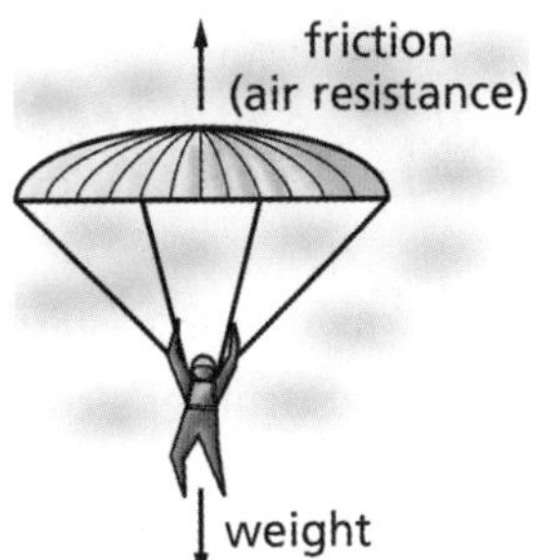

5 An object falling through the air has 2 forces on it: its weight (the force of gravity acting on it), and friction.

The faster it falls, the greater the air resistance.

a When the parachutist starts to fall he is moving slowly, so air resistance is low. The downward force is bigger, so he accelerates downwards.

b The faster he falls, the greater the air resistance.

c Once the air resistance becomes so large that it equals his weight, the forces are balanced, so his speed stays constant – the **terminal velocity** – until he hits the ground.

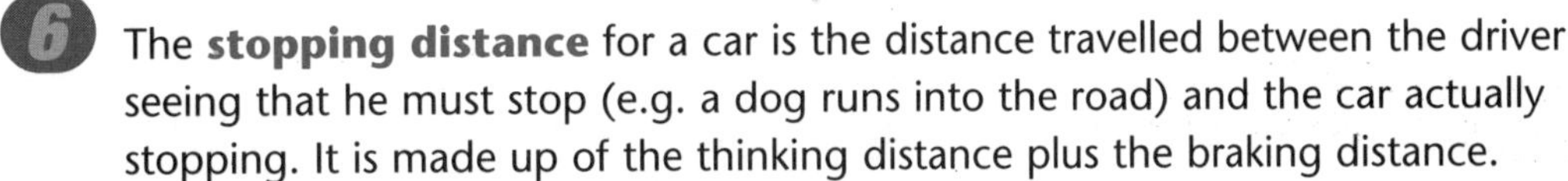

6 The **stopping distance** for a car is the distance travelled between the driver seeing that he must stop (e.g. a dog runs into the road) and the car actually stopping. It is made up of the thinking distance plus the braking distance.

The **thinking distance** is the distance the car moves during the driver's 'reaction time' – i.e. until his foot hits the brakes.

The **braking distance** is the distance moved while the car slows down, due to the force of friction from the brakes. This will be longer the faster the car was moving, the greater the mass of the car, or if the brakes are worn.

Checklist

Are you sure you understand these key terms?

balanced or unbalanced forces / resultant force / reaction force / stopping distance / terminal velocity / thinking and braking distances

Now learn how to use your knowledge

1 A child's plastic football is dropped from the top of a tower block on a summer's day when there is no wind. As it passes the top floor windows the forces on it are as shown below, and its velocity is 5 m/s.

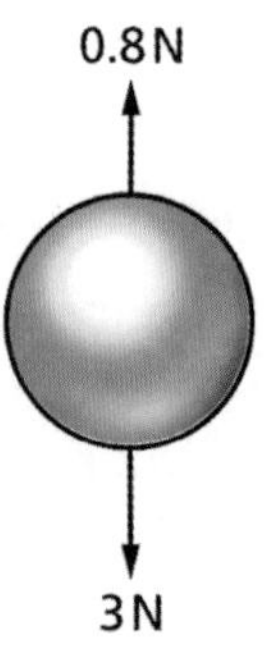

a What is the cause of the downward force on the ball?

....

b What is the cause of the upward force on the ball?

....

c Describe the ball's motion at this point. Hint 1

....

d Lower down, when the ball passes the seventh floor, it is moving at a velocity of 20 m/s. What changes, if any, will there have been to the sizes of Hint 2

the upward force? ....

the downward force? ....

e By the time it has reached the second floor, the ball has reached its terminal velocity.

1) What is meant by terminal velocity?

....

....

2) Explain, in terms of the forces on the ball, why the terminal velocity was reached. Hint 3

....

....

....

2 The minimum stopping distance for Michael's car is usually 75 m when travelling at 60 mph in good weather conditions. However, having loaded the car up ready to go on holiday, the stopping distance will be nearer 80 m.

a Explain why the stopping distance has increased. Hint 4

....

....

b List 2 other factors which affect the stopping distance of a car.

1) ....

2) ....

✔ Hints and answers follow

Forces in action

Hints

1. The forces aren't balanced – there is a resultant force acting downwards. What happens when an object is acted on by an unbalanced force?

2. What happens to air resistance as an object moves faster?

3. If the ball is now moving at a constant speed, are the forces balanced or unbalanced?

4. What is the main difference between an empty car and a full one?

Answers

1 a) weight (or gravity) b) air resistance (or friction) c) the ball is accelerating downwards d) upward force increases, downward force stays the same e) 1) the final, constant speed reached by a falling object when it is no longer accelerating 2) as the ball fell faster the air resistance increased until it equalled the ball's weight. The 2 forces were then balanced so the ball continued with constant speed. **2** a) a fully loaded car has more mass, so the stopping distance is greater b) any 2 of: tired or drunk driver/wet or icy road/worn brakes

Even more forces

Test your knowledge

1 ***a*** A force of 200 N is put on an area of 2 m^2. What is the pressure applied?

b The point of a drawing pin has a tiny cross-sectional area. When you push it into the wall the pressure is .... so the pin goes in easily.

2 A hydraulic system works because liquids can .... pressure.

3 A stretched wire will return to its original size when the force is removed, providing the .... has not been reached.

4 ***a*** The force of gravity acting on an object is called its ....

b The weight of an object acts at its ....

5 Which of these spanners will produce the greatest turning effect?

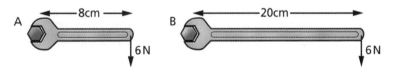

A ←8cm→ 6N

B ←20cm→ 6N

6 As the see-saw isn't moving, the anticlockwise ....
must balance the clockwise ....

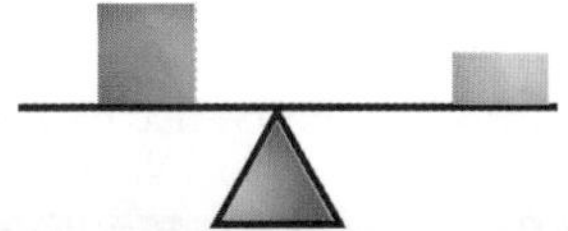

Answers

1 a) 100 N/m^2 b) high **2** transmit **3** elastic limit **4** a) weight b) centre of mass **5** B **6** moments / moments

If you got them all right, skip to page 46

Even more forces

$$\textbf{pressure} = \frac{\text{force}}{\text{area}}$$

So – the *bigger* the force the *greater* the pressure
but the *smaller* the area the *greater* the pressure.

[If the force is in newtons (N) and the area in m^2 then the pressure is in N/m^2 or pascals (Pa).]

Liquids transmit pressure in all directions. Hydraulic brakes on lorries, for example, work on this principle:

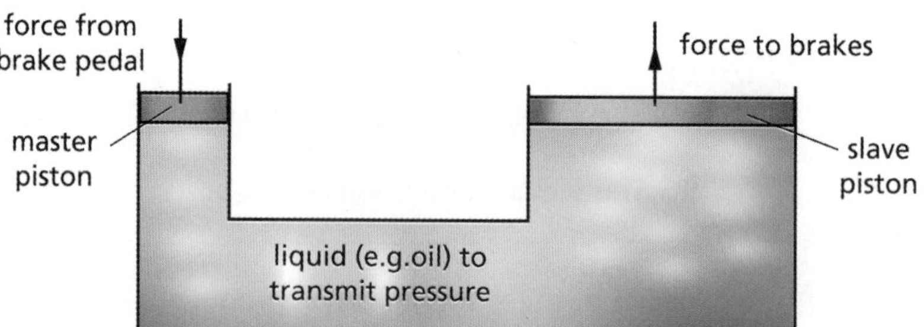

Pushing the brake puts a force on the master piston. Because this piston has a *small* area, the pressure is *high*. This pressure is transmitted through the liquid to the slave piston. This high pressure is now applied over a *large* area, giving a *much greater* force.

So, a small force on the brake pedal puts a large force on the brakes themselves.

When a force is used to stretch a spring (or a metal wire):

- the *greater* the force, the *greater* the spring's extension
- the spring will usually return to its original size when the force is removed.

However, if the spring is stretched so far that the **elastic limit** is passed, the spring will be permanently deformed.

The **weight** of an object is the force of gravity acting on it, in N.
The **mass** of an object is how much matter there is in an object, in kg.
Weight is different on different planets as the gravity varies, but mass is always the same.

The weight of an object acts from its **centre of mass**:

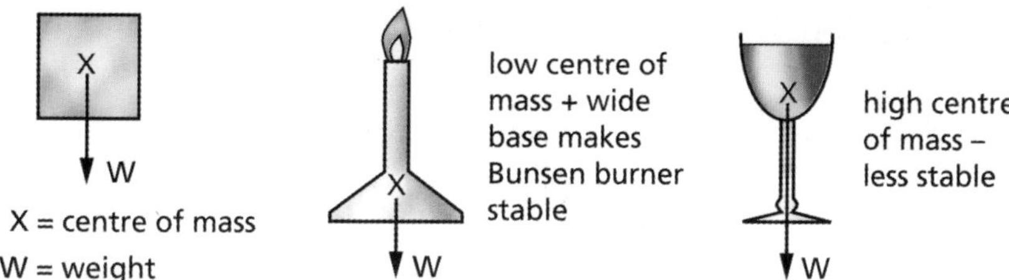

The turning effect of a force will be bigger if:

- the force is bigger
- the force is applied further from the pivot

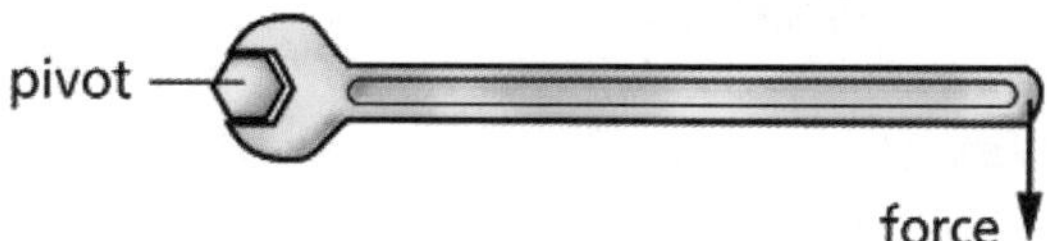

The **moment** of a force tells us how big its turning effect is (in Nm).

moment = force x perpendicular distance from pivot

If an object isn't turning, the *clockwise moments* (all the turning effects trying to turn it clockwise round the pivot) must equal the *anticlockwise* moments.

e.g.

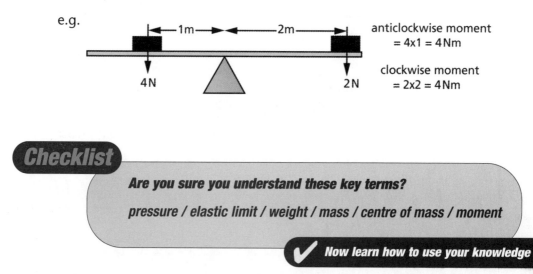

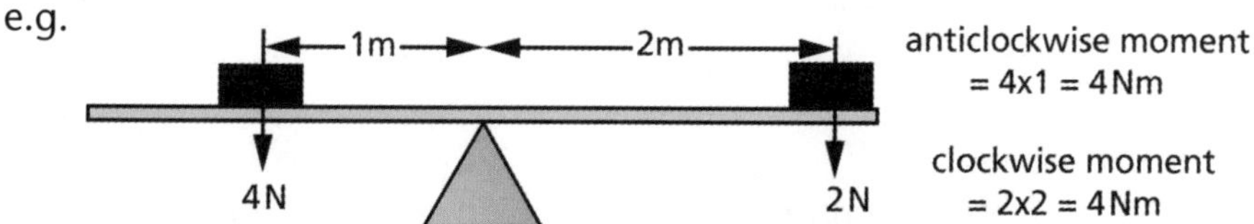

Are you sure you understand these key terms?

pressure / elastic limit / weight / mass / centre of mass / moment

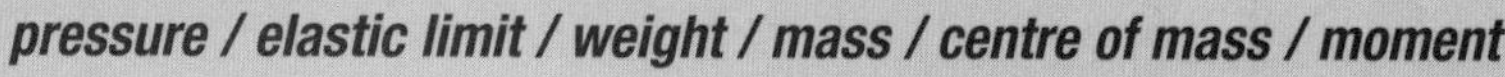

Now learn how to use your knowledge

Even more forces

1 Mark is using a screwdriver to lever the lid off a tin of paint:

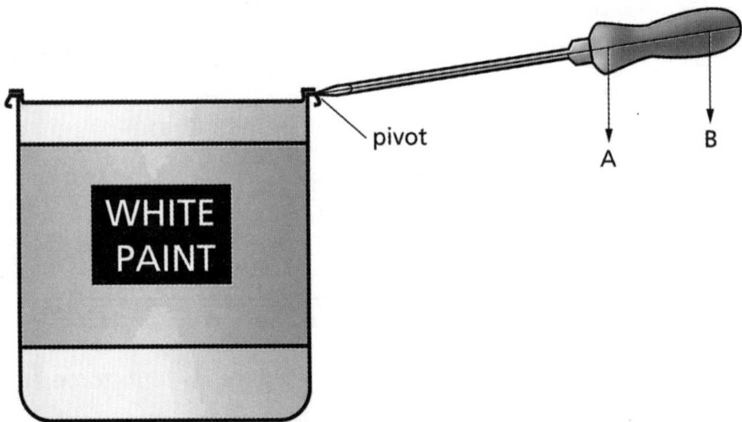

a Would it be better for him to push on the screwdriver at A or B? Why?

Hint 1

....

....

....

b If he pushes on the screwdriver a distance of 0.2 m from the pivot, and the force he applies is 20 N, calculate the moment of the force.

(moment = force x perpendicular distance from pivot)

....

....

....

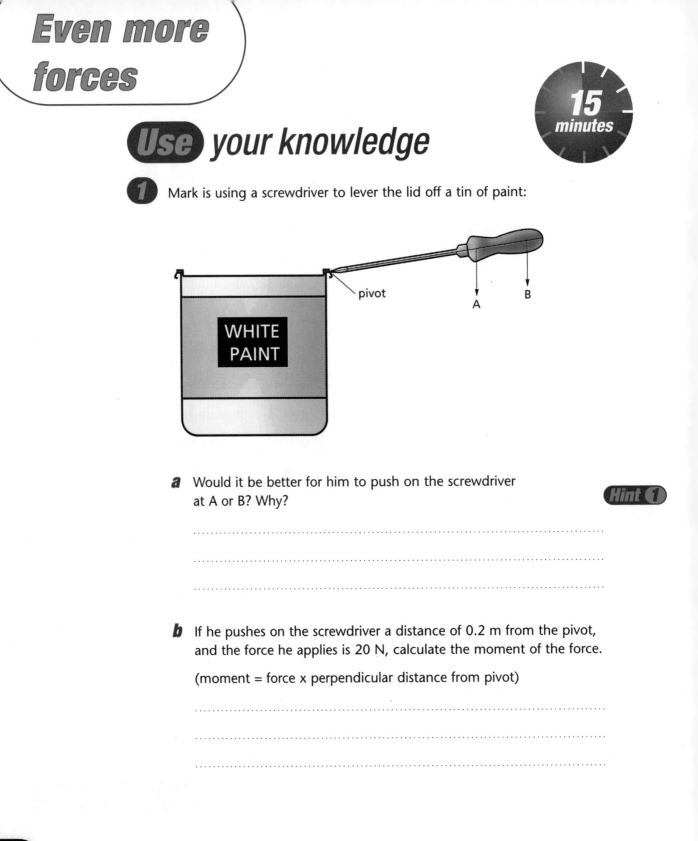

c The paint lid is stuck so tightly that it doesn't move, in spite of Mark's efforts. Explain, in terms of moments, what is happening to the screwdriver. **Hint 2**

....

....

....

....

2 Anita lives in Switzerland. As she rushes out to go to school in the snow she sinks deeply. She then puts on her snowshoes, which have a much wider base than her normal shoes.

Explain why the larger base of the snowshoes will prevent Anita sinking as deeply into the snow. **Hint 3**

....

....

....

....

3 The following graph shows the extension of a metal wire as a force is applied:

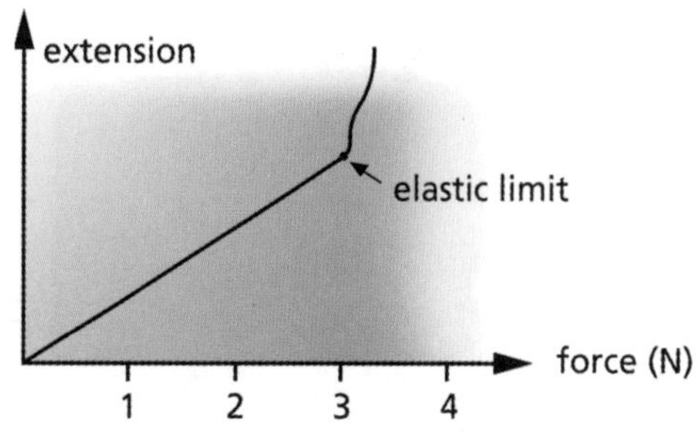

a Describe the shape of the graph up to the elastic limit.

....

b The wire is stretched by a force of 2 N. When the force is removed, what will happen to the wire? **Hint 4**

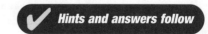

....

Hints and answers follow

Even more forces

Hints

1 He wants to get as big a turning effect as possible.

2 If Mark's force was the only one acting on the screwdriver, it would turn.

What else puts a force on the screwdriver, and so has a 'moment'?

3 Think about the pressure of the shoe on the snow.

What 2 things does pressure depend on?

How does the area of the shoe affect the pressure?

4 Look at the graph – is 2 N more or less than the elastic limit?

Answers

1 a) at B, because the greater the distance from the pivot, the greater the turning effect b) 4 Nm c) the clockwise moments must exactly equal the anticlockwise moments, as it isn't turning (the anticlockwise moment is caused by the force of the lid on the tip of the screwdriver)

2 the snowshoes spread Anita's weight over a larger area, so putting less pressure on the snow

3 a) a straight line b) the wire will return to its original length (because the elastic limit has not been exceeded)

Radioactivity

Test your knowledge

1 Alpha, beta, or gamma radiation is emitted when the .... of an atom decays.

2 *a* The most penetrating type of radiation is .... radiation.

b The least penetrating type of radiation is .... radiation.

c The type of radiation which will pass through a sheet of paper but is absorbed by 5 mm of metal is .... radiation.

3 The low level of radiation which is around us all the time is called .... radiation.

4 The level of radiation can be found using a ....

5 What effects can radiation have on the cells in our body?

6 *a* In what way can radiation be used in the treatment of cancer?

b What is the name of the radioactive isotope used to date fossil remains?

Answers

1 nucleus **2** a) gamma b) alpha c) beta **3** background **4** Geiger counter (or Geiger-Muller tube) **5** it ionises them, causing damage, cancer, or cell death **6** a) to kill cancerous cells (radiotherapy treatment) b) carbon-14

✓ If you got them all right, skip to page 52

Radioactivity

Improve your knowledge

1 Radioactive emission occurs as a result of changes to the **nucleus** of an atom. Some types of atom have unstable nuclei, which try to become more stable by **decaying**. They emit radiation, and form a different type of atom.

2 **Alpha radiation** (α) can be stopped by a sheet of paper or a few centimetres of air. It consists of helium nuclei.

Beta radiation (β) is not stopped by air or paper, but most of it is stopped by a few millimetres of metal (e.g. aluminium foil). It consists of electrons.

Gamma radiation (γ) is the most penetrating of the three. Several centimetres of lead, or several metres of concrete will absorb most of it. Gamma radiation is part of the electromagnetic spectrum, with a very high frequency.

3 There is a small amount of radiation around us all the time, called **background radiation**. This is caused by radioactive substances in the Earth itself, in building materials, in the air, in food, even in our bodies, and from space. Some areas of the country have higher levels of background radiation, due to the presence of certain rocks, such as granite.

4 Radiation can be detected using a **Geiger counter**. It is usually measured in **counts per minute** – the number of radioactive emissions entering the counter in one minute. Before measuring the radioactivity from a sample, the level of background radiation must be recorded, so that it can be taken into account in the measurements.

5 Radiation, especially in large amounts, can be dangerous to humans. This is because when radiation collides with neutral molecules it can cause them to become charged or **ionised**. If the molecules in our body cells are ionised, they may be damaged, or even turn cancerous. Larger doses of radiation kill cells.

Precautions must be taken when handling radioactive substances. In the lab, samples are handled at arm's length, using tweezers and wearing gloves and glasses. People who work with radioactive materials wear a badge containing photographic film. This is developed to show the level of radiation they have been exposed to, to check that they do not exceed the safety limits.

6 Radioactive substances have a variety of uses:

a *Radiotherapy* uses radiation to kill cancer cells.

b *Radioactive tracers* can be introduced into the body, and their path through the body detected from outside. This enables certain medical problems to be diagnosed without an operation.

c *Thickness gauges* used in the production of paper consist of a source of beta radiation above and a detector below the paper. The thicker the paper, the more radiation is absorbed, so less reaches the detector.

d *Carbon–14* dating is used to estimate the age of fossils. Living bodies contain a lot of carbon, of which a constant fraction is the radioactive isotope, *carbon–14.* Once the body dies, the amount of *carbon–14* drops slowly as the atoms decay radioactively and are not replaced. So measuring the fraction of *carbon–14* remaining allows the age of the remains to be judged.

Checklist

Are you sure you understand these key terms?

nucleus / decay / alpha, beta and gamma radiation / Geiger counter / background radiation / counts per minute / ionised

✔ ***Now learn how to use your knowledge***

Radioactivity

An experiment to investigate a radioactive source is set up as shown below:

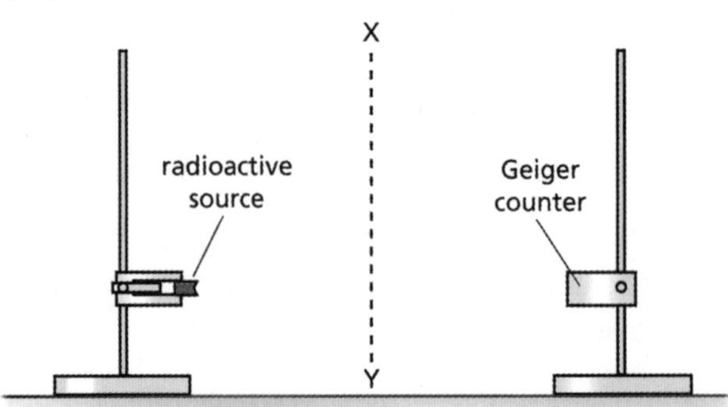

Different materials, such as paper and foil, may be put between the source and the counter, in the position marked X–Y.

The following readings were taken:

	Counts per minute
with no source present	6
with source, as shown in the diagram	98
with source, and a sheet of paper clamped between X and Y	36
with source, and a 3 mm thick sheet of aluminium clamped between X and Y	36

1 With no radioactive source present, the Geiger counter does not read zero. Why not?

....

....

....

2 What is the radioactivity, in counts per minute, due to the source alone? **Hint 2**

....

3 Some of the radiation is stopped by a single sheet of paper. What type of radiation is this likely to be? **Hint 3**

....

4 No further radiation is cut out by the aluminium sheet. However, the count rate is still significantly above the original reading of 6 counts per minute. What other type of radiation does this suggest might be present? **Hint 4**

....

5 Give 2 precautions that should be taken when carrying out this experiment.

a ....

b ....

6 Why are precautions needed when dealing with radioactive substances?

....

....

....

7 List 2 uses of radioactive materials.

a ....

b ....

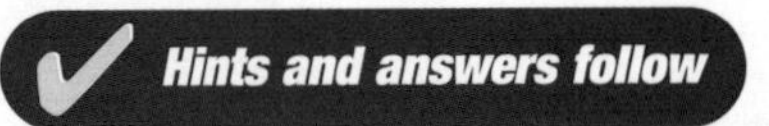

Radioactivity

Hints

1 There is always some radiation present around us.
What is it called? What causes it?

2 The 98 counts per minute is due to the source *plus* the background radiation.

The background radiation count is the reading given before the source was put in place.

So, the radioactivity due to the source is 98 minus 6 counts per minute, which is 92 counts per minute.

3 Only one type of radiation can be absorbed by a sheet of paper. Which one?

4 Both alpha and beta radiation would be cut out by this much aluminium. But some radiation is getting through the sheet. So what kind of radiation must be present?

(Notice that adding the aluminium sheet didn't make any difference to the count rate. This means that the source can't be producing any beta radiation, as this would have mostly gone through the paper, but would then be stopped when the aluminium was used.)

Answers

1 because of the background radiation from rocks, the air, our bodies, etc. **2** 92 counts per minute **3** alpha radiation **4** gamma radiation **5** a) e.g. handle at arm's length b) use tweezers **6** radiation can damage human cells and cause cancer **7** a) (e.g.) radiotherapy for cancer treatment b) carbon–14 dating of fossils

Circuits

Test your knowledge

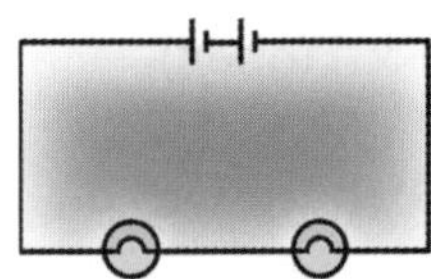

1. The circuit above contains 2 .... and a ....

2. An electric current is a flow of ....

3. The greater the resistance of a circuit, the .... the current.

4. In the circuit above, electrical energy is transferred to .... energy.

5. In the circuit above, A and B are in .... with each other.

6. **a** The current through A could be measured using an ....

   b To measure the voltage across A, a voltmeter should be placed in .... with A.

7. Would a current–voltage graph for A be a straight line or a curve?

   ....

Answers

1 bulbs / cell **2** electrons (or charge)
3 smaller **4** light (and heat) **5** series
6 a) ammeter b) parallel **7** curve

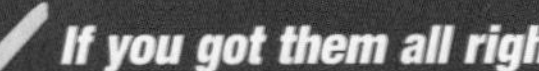
If you got them all right, skip to page 58

Circuits

1 The main symbols used in circuit diagrams are:

- switch (open)
- switch (closed)
- resistor
- variable resistor
- diode (current can only travel in the direction of the 'arrow' – in this case to the left)
- cell (2 or more make a battery)
- ammeter
- voltmeter
- lamp

2 An **electric current** is a flow of charge – usually a flow of electrons.

A current will flow if there is a **voltage** provided (e.g. by a battery) and a circuit for it to flow around. The higher the voltage the greater the current. (*Voltage* is sometimes called **potential difference**.)

3 **Resistance** makes it harder for the current to flow in a circuit. The greater the resistance is, the smaller the current will be.

When a large current flows through a resistor, the resistor is heated. This effect is used in heating appliances, e.g. hair dryers, and is why light bulbs get hot.

voltage = current x resistance **Learn this**

Voltage is measured in volts (V), current in amps (A), and resistance in ohms (Ω).

4 In an electric circuit, energy is transferred (or converted) from electrical energy in the battery to other types of energy in the components (heat in resistors, light in bulbs, movement in motors, etc.)

The **power** of a circuit is the rate at which this energy is transferred (in watts, W). We can calculate it:

power = current x voltage **Learn this**

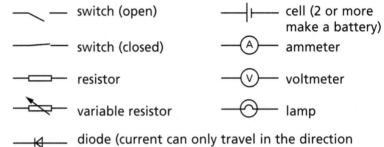

5 Electric circuits can be connected in **series** or **parallel**:

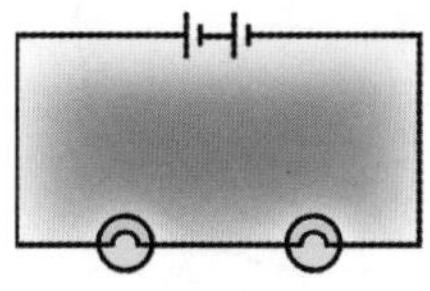

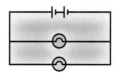

2 lamps in series — 2 lamps in parallel

6 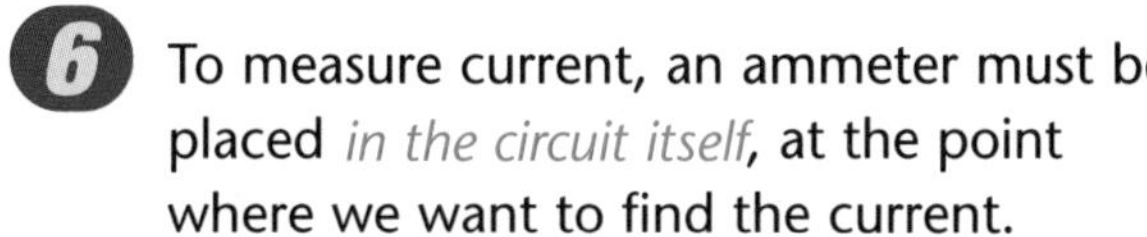

To measure current, an ammeter must be placed *in the circuit itself*, at the point where we want to find the current.

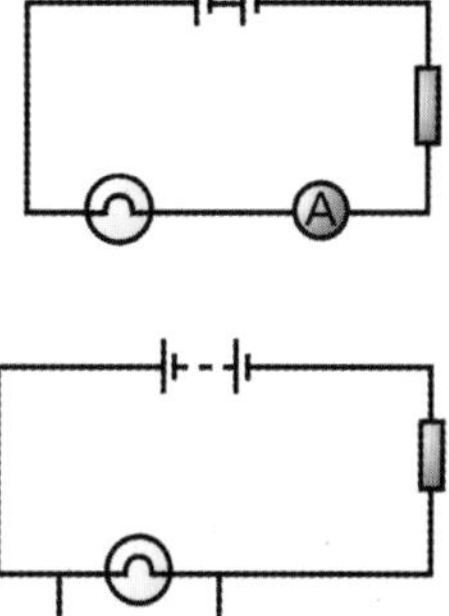

To measure voltage, a voltmeter must be placed *in parallel with* the component we are interested in.

7 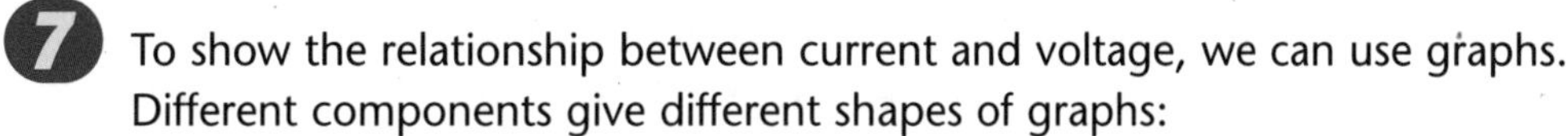

To show the relationship between current and voltage, we can use graphs. Different components give different shapes of graphs:

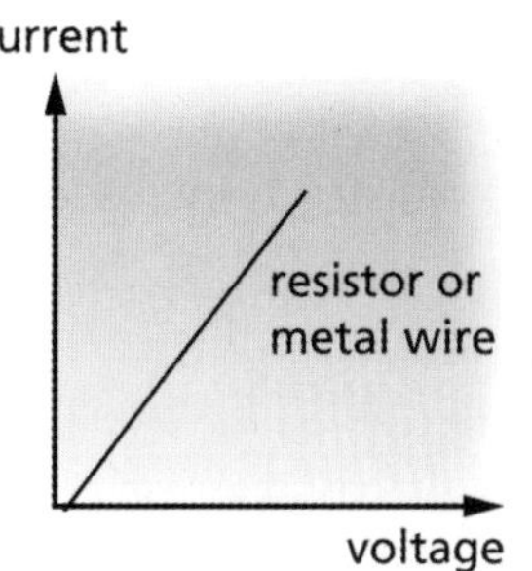

a straight line shows resistance is constant – (*if* it doesn't heat up!)

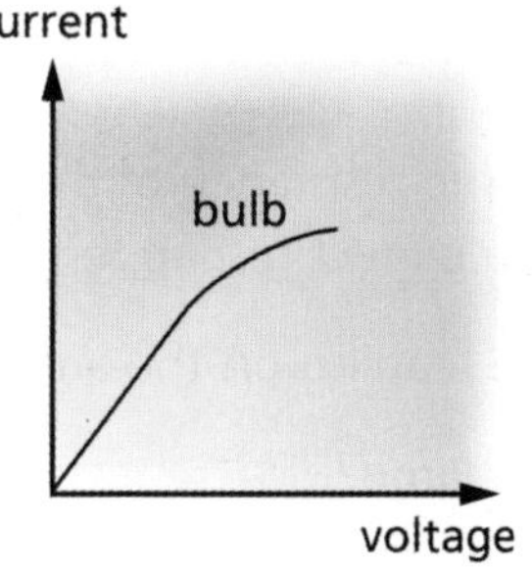

as the bulb gets hotter its resistance increases, and the line curves

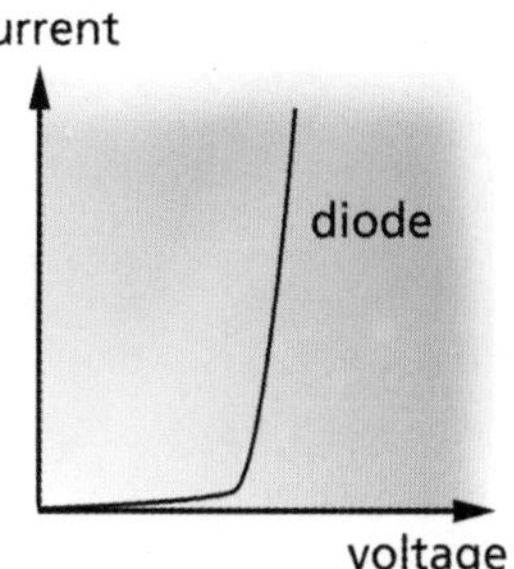

once a certain voltage is reached, a large current flows

Checklist

Are you sure you understand these key terms?

electric current / voltage / potential difference / resistance / power / series circuit / parallel circuit

 ***Now learn how to use your knowledge***

Circuits

Use your knowledge

1 The diagram shows a simple circuit diagram:

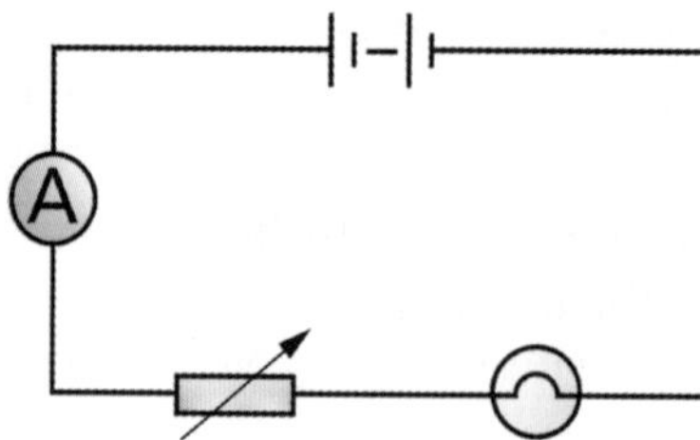

a Will the current through the lamp be larger, smaller, or the same as the current through the variable resistor? **Hint 1**

....

b As the resistance of the variable resistor is increased, what would you expect to happen to **Hint 2**

the reading on the ammeter? ....

the lamp? ....

c Now add a suitable component to the diagram, to measure the voltage across the variable resistor. **Hint 3**

d Describe the type(s) of energy transfer taking place in the circuit. **Hint 4**

....

....

....

e The bulb has printed on it '3 V; 0.5 A'. If the bulb was operating to those specifications, calculate the power consumed by the bulb.

....

....

....

2 The graph shows how the current through a diode varies when the voltage is changed.

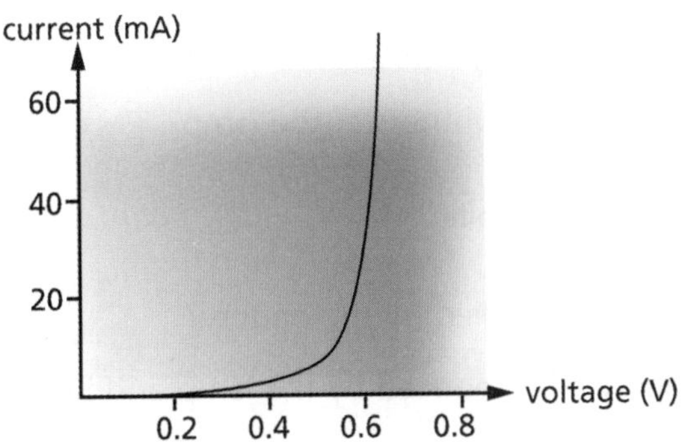

a Describe in words what happens to the current through the diode as the voltage is increased.

....

....

....

b What would happen if the voltage was negative – i.e. the voltage supply was connected the other way round?

....

....

Hints and answers follow

Circuits

Hints

1 The variable resistor and lamp are in series. All the current flowing through the variable resistor must also flow through the lamp, as there is nowhere else for it to go to.

2 What effect does increasing the resistance have on the current?

What will happen to the brightness of the lamp?

3 Does this go in series or parallel?

4 What kind of energy is supplied by the battery?

What kinds of energy are produced by the lamp and by the resistor?

5 You must learn the equation for power.

6 What happens when the voltage is small?

What happens at a larger voltage?

7 It matters which way round a diode is connected into a circuit – why?

What happens if a diode is connected the wrong way round?

Answers

1 a) the same as b) ammeter reading decreases, bulb gets dimmer c) see diagram d) electrical energy (from battery) is transferred to light energy (in bulb) and heat (in bulb and resistor) e) 1.5 W **2** a) below 0.6 V there is a tiny current, but at around 0.6 V the current increases rapidly b) no current flows (current can only flow one way through a diode)

1 c)

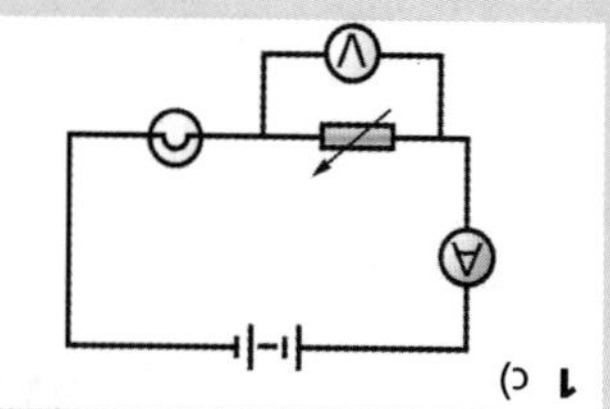

Electricity at home

Test your knowledge

1 Which wire in a 3-pin plug is connected to the fuse?

2 Which wire does not usually have any current flowing through it?

3 If the metal case of an appliance becomes 'live', the current is carried safely away through the .... wire.

4 ***a*** A circuit breaker will switch off a circuit if the current in the .... wire is different from the current in the .... wire.

b You have the following fuses: 1 A, 3 A, 5 A, 13 A.

Which fuses should be used in an appliance where the current is

1) 4 A   2) 6 A?

5 energy = power x time

How much energy (in joules) does a 1500 W heater use

a in 1 second?

b in 5 minutes?

6 ***a*** The current used in our homes is called .... current, because it changes direction many times a minute.

b Batteries, however, provide .... current.

Answers

1 live **2** earth **3** earth **4** a) live / neutral
b) 1) 5 A 2) 13 A **5** a) 1500 J b) 450,000 J
(remember 5 minutes is 5 x 60 seconds)
6 a) alternating b) direct

If you got them all right, skip to page 64

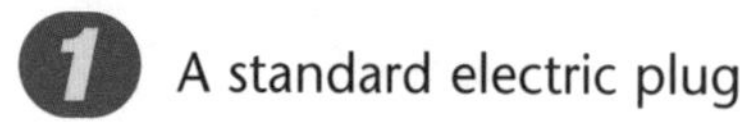

Improve your knowledge

1 A standard electric plug:

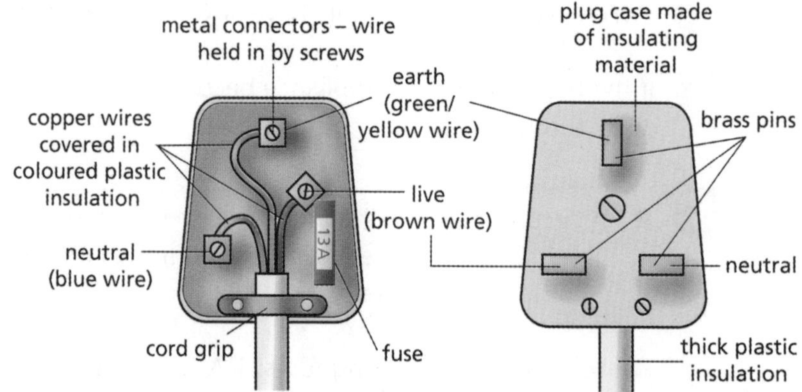

2 The electric current is supplied to the appliance through the **live** wire. This is at around 230 V, and could kill you if touched.

The current leaves the appliance through the **neutral** wire (0 volts).

The **earth** wire is provided for safety. In normal use, no current flows through it.

3 Appliances with a metal case should be **earthed**. A connection is made between the case and the earth wire. If the case became 'live', the current would be safely carried to earth. This large flow of current would also blow the fuse, turning off the appliance.

Without an earth wire, someone touching the 'live' case could get a serious electric shock as the current flowed through them to the earth.

Appliances with an insulating case (e.g. plastic), in addition to the insulation on the electric wires are **double insulated**. They do not require an earth wire.

4 The current has to go through the **fuse** between the live pin and the live wire. If too much current flows, the thin piece of wire inside the fuse melts, and the current is stopped, e.g. a 13 A fuse melts if a current of 13 A or more flows through it.

Electricity enters the house through a fuse box, containing a series of either fuses or circuit breakers. A **circuit breaker** will switch off the current *if* the current in the live wire and neutral wire isn't the same (due to a surge or leak in current). They are more sensitive than fuses.

5 energy = power x time *you must be able to use this*

- If power is in watts (W) and time in seconds (s) then the energy is in joules (J). *(used in the lab)*
- In calculating energy used in the home, different units are used: if power is in kilowatts (kW) and time in hours (h), then energy is in **kilowatt-hours** (kWh).

Electricity bills charge us for the number of kilowatt-hours of electricity we have used. They call a kilowatt-hour **one unit** of electricity.

The 'unit price' is the price for each kilowatt–hour used.

6

- The electric current supplied by a battery is **direct current (d.c.)** so current always flows in the same direction.
- The electric current supplied by the mains to our homes is **alternating current (a.c.)**. The current is constantly changing direction. In the UK our mains electricity is 50 Hz a.c., which means the direction of the current changes over and back again 50 times per second!

Checklist

Are you sure you understand these key terms?

live / neutral / earth / double insulated / fuse / circuit breaker / kilowatt-hour / unit / alternating current / direct current

Now learn how to use your knowledge

Mary has a stainless steel electric kettle. Its plug is wired as shown below:

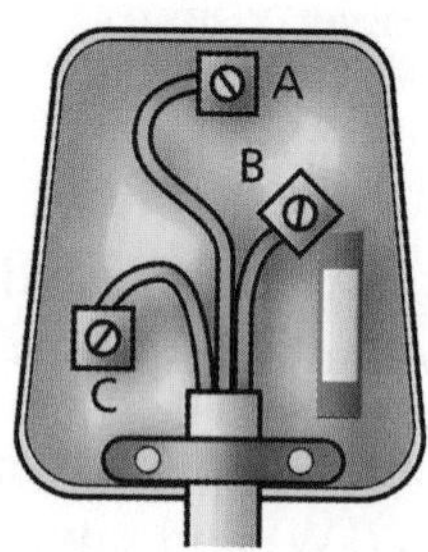

1 What is the name of the wire attached to A?

....

2 What is the purpose of this wire? Hint 1

....

....

....

Mary's kettle is old and a bit temperamental, so she decides to buy a new, smarter, plastic kettle.

On starting to wire the plug, Mary is dismayed to notice that her new kettle has only 2 wires. The wire that would be connected to pin A is missing.

3 Explain why this wire is not necessary for her new kettle. Hint 2

....

....

....

Mary's new kettle has the following label attached to it:

The instructions with the kettle state that a 13 A fuse should be used.

4 What would happen if Mary used a 5 A fuse instead? **Hint 3**

....

....

5 If Mary has her kettle switched on for a total of 3 minutes a day, calculate the amount of energy used, per day, in joules. **Hint 4**

(energy = power x time)

....

....

6 In 3 months, Mary's kettle will have been used for about $4^1/_2$ hours. Calculate the amount of energy used in that time, in kilowatt-hours. **Hint 5**

....

....

7 If the price of electricity is 6 pence per unit, calculate the cost of using Mary's kettle over that 3-month period. **Hint 6**

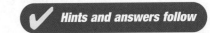

....

....

✔ Hints and answers follow

Electricity at home

Hints

1 Explain what happens if the metal case becomes live.

2 The new kettle does not have a metal case.

3 A 5 A fuse melts if a current of 5 A or more flows through it.

4 The time must be in seconds to get an answer in joules.

5 a) Use the same equation as for question **4**.

b) To get an answer in kilowatt-hours you must have the power in kilowatts and the time in hours.

c) 2300 W is 2300 ÷ 1000 kW.

6 a) 6 p per unit means 6 p for each kilowatt-hour.

b) So, multiply the number of kilowatt-hours by 6 p to get the total price.

If you weren't sure how to do questions **6** and **7**, look at them again carefully. Make sure you could do similar questions again.

Answers

1 earth **2** if the case becomes live, the earth wire carries the current safely to earth **3** it has a plastic case so is double insulated **4** the fuse would melt ('blow') at once **5** energy = power x time = 2300 x 180 = 414,000 J **6** energy = power x time = 2.3 x $4\frac{1}{2}$ = 10.35 kWh **7** cost = 10.35 x 6p = 62.1p

Electromagnetism

Test your knowledge

1 Will the following pairs of magnets be attracted, repelled, or unaffected by each other?

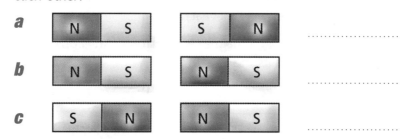

a N S S N ....

b N S N S ....

c S N N S ....

2 ***a*** A coil of wire will behave as a magnet if a .... is flowing through it.

b A .... bar through the centre of the coil will increase the strength of the magnet.

3 What happens to a wire with an electric current going through it, if it is put in a magnetic field?

4 What happens to a metal wire moving in a magnetic field?

5 ***a*** What device is used to increase the voltage of electricity leaving a power station?

b Why is it better to send electricity round the country at a high voltage?

Answers

1 a) repelled b) attracted c) repelled **2** a) current b) soft iron **3** it has a force on it **4** a voltage is induced across it **5** a) transformer b) a low current can be used, so less energy is wasted as heat in the wires.

If you got them all right, skip to page 70

Improve your knowledge

1 **Bar magnets** have a north pole and a south pole. The north pole of one magnet will attract the south pole of another magnet, but two north poles or two south poles will repel each other.

The space around a magnet where iron or steel objects will feel a force is called its **magnetic field**.

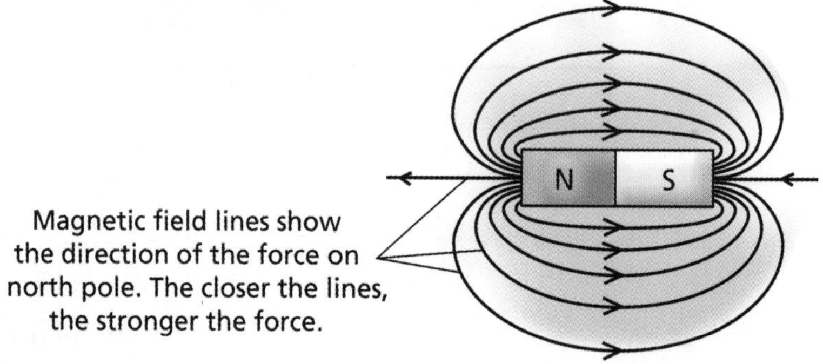

N
S
Magnetic field lines show the direction of the force on a north pole. The closer the lines, the stronger the force.

2 An **electromagnet** is just a coil of wire with an electric current going through it. It behaves just like a bar magnet when the current is switched on.

The strength of the electromagnet can be increased by

- ***a*** increasing the current
- ***b*** using a coil with more turns of wire
- ***c*** putting a soft iron core through the centre of the coil.

Electromagnets are often more useful than permanent bar magnets, as they can be turned on and off.

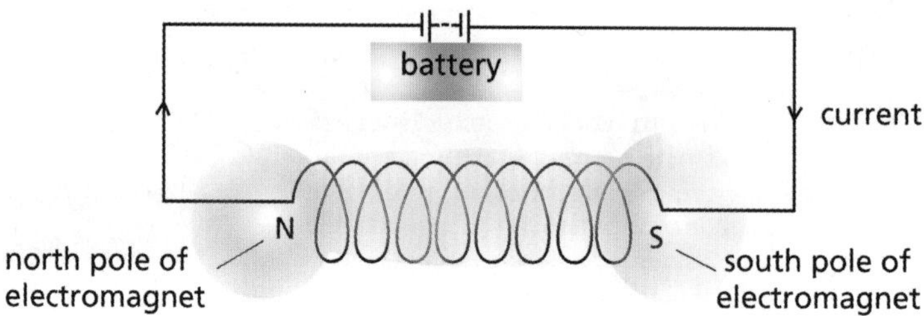

3 If a wire with current flowing through it is put in a magnetic field, it will experience a force (unless it is parallel to the magnetic field lines).

In an electric motor, one side of the coil has a downward force on it. The other side has an upward force, as the current is flowing in the opposite direction. These forces make the coil rotate.

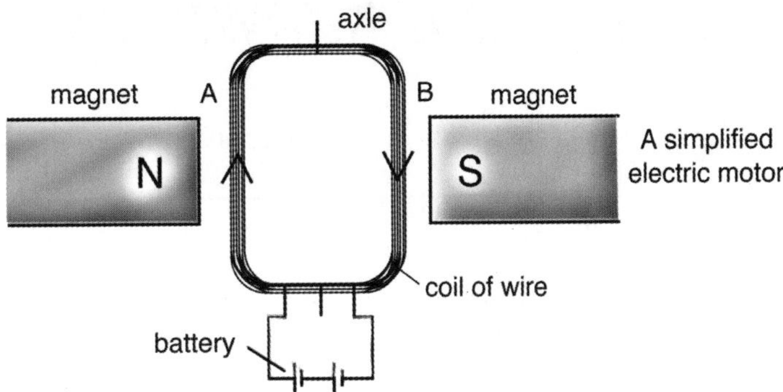

A simplified electric motor

4 A conductor (e.g. a metal wire) will have a voltage **induced** across it if

a the magnetic field around it is changing

b it is moving through a magnetic field.

If the conductor is part of a complete circuit, then the induced voltage causes a current to flow in the circuit.

This principle is used to generate electricity in power stations, when a magnet rotates in a coil of wire.

5 **Transformers** are used to change the size of an a.c. voltage. At the power station they increase the voltage of the electricity before it is sent around the country on power lines. The voltage is reduced again by local transformers to make it safe for use in homes.

High voltages are used in the power lines because then only a low current is needed. This causes much less heating in the cables, so less energy is lost as heat.

Checklist

Are you sure you understand these key terms?

bar magnet / magnetic field / electromagnet / induced voltage / transformer

✔ ***Now learn how to use your knowledge***

Electromagnetism

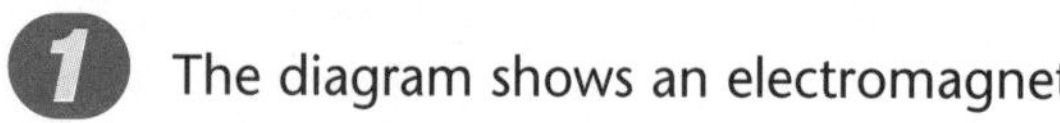

Use your knowledge

1 The diagram shows an electromagnet:

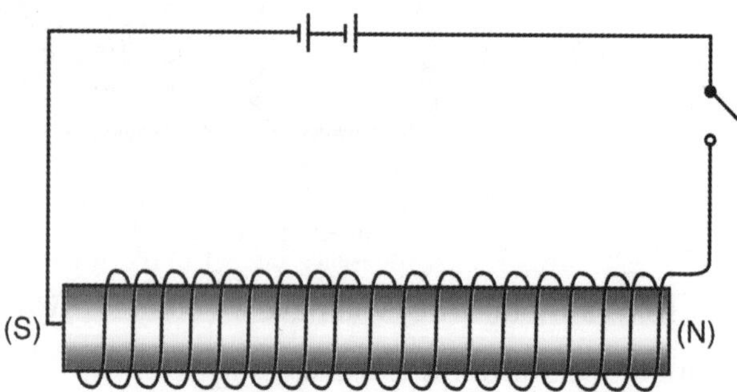

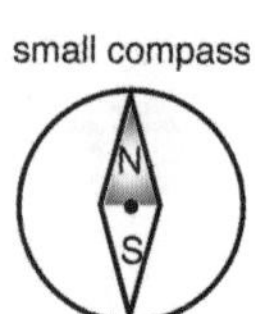

When the electromagnet is switched on it has North and South poles, marked (N) and (S).

The compass pointer is a small permanent magnet, with poles marked N and S.

a Describe what happens to the compass when the electromagnet is turned on.

....

....

b On the diagram, draw the shape of the magnetic field around the electromagnet.

2 The diagram below shows a relay. It allows a tiny current from an electronic circuit to switch on a much larger current powering a motor.

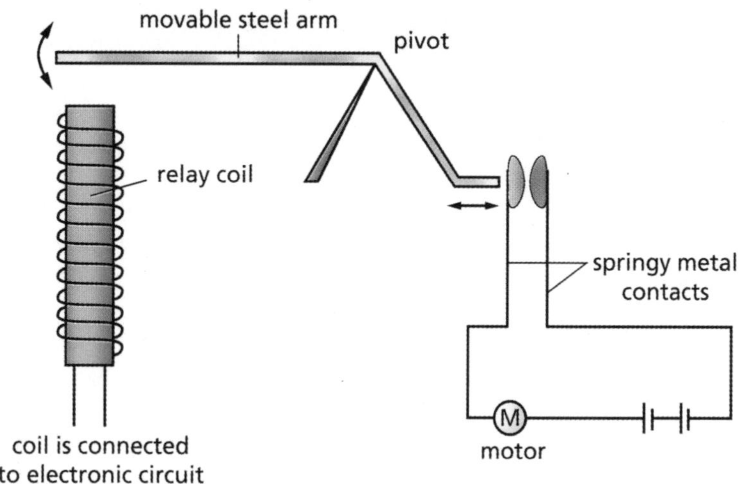

When the relay coil is switched on by the electronic control circuit, describe what happens to

a the relay coil .... **Hint 3**

....

b the movable arm .... **Hint 4**

....

....

c the metal contacts .... **Hint 5**

....

....

d the motor .... **Hint 6**

....

....

✔ Hints and answers follow

Electromagnetism

Hints

1 The compass is close to the north pole of the electromagnet, and the pointer is free to turn around.

Which end of the compass pointer is attracted to the north pole of the electromagnet?

2 Remember a coil of wire behaves as a bar magnet when it has a current going through it.

So, the magnetic field will be the same shape as for a bar magnet.

3 What happens to a coil of wire when a current flows through it?

4 The arm is made of steel, which is attracted by magnets.

Notice that the arm has a pivot, around which it can turn.

5 As the arm pivots, the right hand side of it will push against the contacts.

6 With the contacts touching each other, there is now a complete electric circuit with the motor in.

So, the contacts act as a kind of switch in the motor circuit.

Answers

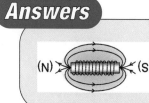

1 a) the south pole of the pointer swings round to point towards the north pole of the electromagnet b) see diagram **2** a) it becomes a magnet b) the left side is attracted to the electromagnet, making the arm pivot c) the arm pushes the contacts together so they are touching d) the motor is switched on

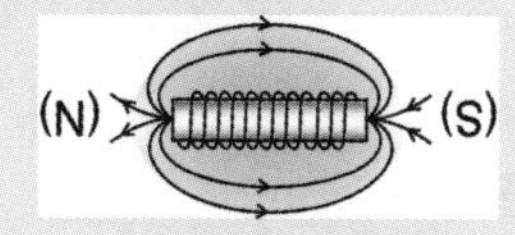

Charge

Test your knowledge

1 When two materials are rubbed together (e.g. a plastic ruler on a wool jumper), they may become charged. This is because .... have been rubbed off one material and onto the other material.

2 A negatively charged rod will .... another negatively charged rod.

A negatively charged rod will .... a positively charged rod.

A negatively charged rod will .... tiny pieces of paper.

3 In electrostatic paint spraying, positively charged paint droplets are sprayed onto a car body. For maximum efficiency, should the car body be:

- ***a*** earthed?
- ***b*** negatively charged?
- ***c*** positively charged?
- ***d*** neutral?

4 ***a*** A spark occurs when .... jumps a gap from one object to another.

b A charged person may become discharged by touching an earthed metal object. The person may feel an ....

5 To prevent a build up of charge which could cause a spark, should fuel bottles be made of

- ***a*** a type of plastic which doesn't conduct electricity?
- ***b*** a type of rubber which doesn't conduct electricity?
- ***c*** a type of plastic which does conduct electricity?

Answers

1 electrons **2** repel / attract / attract **3** b)
4 a) charge (or electrons) b) electric shock
5 c)

If you got them all right, skip to page 76

Charge

Improve your knowledge

1. When 2 non-conducting objects (e.g. plastic, cloth, etc.) are rubbed against each other, they may become **charged** electrostatically as a result of the friction between them. Electrons have been rubbed *off* one object (which becomes *positively* charged) and rubbed onto the other object (which becomes *negatively* charged).

2. 'Like charges repel; opposites attract. So:

   - 2 positively charged objects repel one another,
   - 2 negatively charged objects repel one another,
   - but a negative object attracts a positive object.

   A charged object (positive or negative) will also attract small, light, uncharged objects – such as dust or tiny pieces of paper.

3. One use of electrostatic charge is in the reduction of pollution from power stations. Charged 'dust precipitators' fitted to power station chimneys attract particles of ash and so prevent them being released into the atmosphere.

   A photocopier works by producing a charged pattern on its internal drum to match the original sheet to be copied. The toner (black 'ink') is a powder, and is attracted to the charged areas only. This black pattern is then pressed against a sheet of charged paper, so producing a copy of the original sheet.

Charged objects can be **discharged** (lose their charge) in a number of ways, generally involving either

- being connected to the earth by a conductor
- charge transferring from a negatively to a positively charged object.

If there is a sufficiently large amount of charge, electrons may jump a small gap between 2 objects, giving a visible spark. A large-scale version of this is lightning, which is a huge spark between a charged cloud and another cloud or the earth.

People can become charged in shops, due to friction between their shoes and the floor. Touching an earthed metal conductor (e.g. a clothes rack) may discharge the person, who may feel a slight electric shock.

Remember that electric current is a flow of charge. A movement of charge such as a spark is therefore a brief electric current.

In some situations, a build-up of charge can be dangerous. When plastic bottles for petrol were first introduced there were a number of accidents. Friction between the lid and the bottle could cause a build-up of charge leading to a spark which ignited the petrol. (Modern fuel bottles are made of a type of plastic which conducts electricity, so charge cannot build up in that way.)

In situations where a build-up of charge must be avoided, such as in aircraft refuelling, or in factories dealing with flammable powders, precautions must be taken. Conducting materials are used whenever possible (types of rubber and plastic which conduct electricity are available) and appliances can also be earthed, to carry the charge away safely.

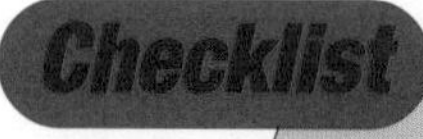

Are you sure you understand these key terms?

charged / discharged

Now learn how to use your knowledge

Charge

Use your knowledge

1 Sasha is combing her hair. Unfortunately as the comb rubs against her hair, electrons are rubbed off the comb onto her hair. Her hair becomes negatively charged.

a Does Sasha's comb become positively or negatively charged? Hint 1

....

b Has Sasha's comb gained or lost electrons? Hint 2

....

c Now that Sasha's hair is charged, she finds it will not lie flat on her head, but that some hairs are drifting away from the others. Hint 3

Explain why Sasha's hair will not lie flat.

....

....

....

....

d If Sasha holds her comb close to her hair, what would you expect to happen? Why? Hint 4

....

....

....

....

A man walking on a non-conducting floor surface may become positively charged, as shown in the diagram:

a What causes him to become charged?

....

....

b What might the man feel if he touched an earthed metal object? Hint 5

....

c Why is it important that the floor of a factory producing cartons of a flammable substance is made of a *conducting material*? Hint 6

....

....

....

Hints and answers follow

Charge

Hints

1 When 2 objects become charged by rubbing against each other, one object will become positively charged and the other negatively charged.

The question tells us that her hair has become negatively charged.

2 The question tells us that her hair has gained electrons from the comb.

(In general, it helps to remember that electrons are negatively charged. So an object becomes negatively charged if it has gained extra electrons. The other object has become positively charged and has lost the electrons.)

3 Sasha's hair is negatively charged. This means that the individual hairs are negatively charged.

What will one negatively charged hair do to another negatively charged hair?

4 The comb and her hair have opposite charges.

5 Touching an earthed object should discharge the man.

6 When an object is discharged, there is usually a spark.

Answers

1 a) positively charged b) lost electrons c) Sasha's hairs are each negatively charged and so will repel one another, and not lie flat against each other d) the hair is attracted to her comb, because the hair and comb have opposite charges **2** a) friction between his shoes and the floor (which causes electrons to be rubbed from his shoes onto the floor) b) an electric shock c) the tiny spark produced when the man or floor was discharged could ignite the flammable substance causing an explosion and/or fire, but charge will not build up on a conducting floor.

Electronic control

Test your knowledge

1 In general, an electronic control system contains an input sensor, a processor, and an .... device.

2 *a* Under what conditions will an LDR switch something on?

b Under what conditions will a thermistor switch something on?

3 *a* Under what conditions will the output of an AND gate be on?

b Under what conditions will the output of a NOT gate be on?

4 Complete the following truth table for an OR gate:

Input 1	Input 2	Output
0	0	
0	1	
1	0	
1	1	

5 A .... can be connected to the output of a processor, to switch on a more powerful circuit containing a heater.

Answers

1 output **2** a) if the light is bright enough b) if it is hot enough **3** a) if both inputs are on b) if the input is not on **4** outputs: 0, 1, 1, 1 **5** relay

If you got them all right, skip to page 82

Electronic control

Electronic control systems consist of three parts

a input sensors

b a processor, or 'control circuit'

c an output device.

The **input sensors** detect conditions in the environment (e.g. temperature, light) and depending on these conditions turn the inputs to the processor on or off. As a result of these inputs, the **processor** switches the **output device** on or off.

2 Input sensors include:

a A **light dependent resistor (LDR)** to detect light. It switches the input ON if the light is bright enough.

b A **thermistor** to detect heat. It switches the input ON if it is hot enough.

c A **switch**. If it is switched on, the input will be ON. Switches can be controlled manually by people, or some switches can be switched on by pressure, by tilting, or by a magnetic field.

3 Processors (control circuits) are usually made up of one or more logic gates. The output of a logic gate depends on its inputs, and on what type of gate is being used.

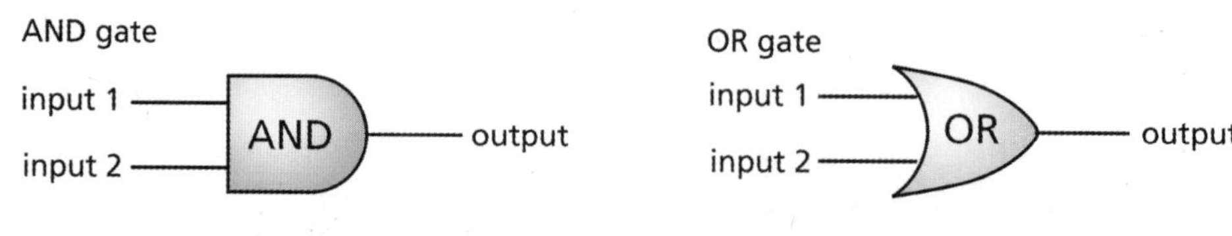

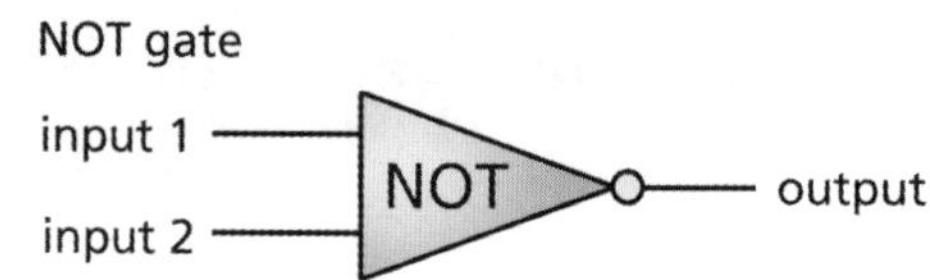

For an **AND gate**: The output is switched on if input 1 *AND* input 2 are on.

For an **OR gate**: The output is switched on if input 1 *OR* input 2 (or both) is on.

For a **NOT gate**: The output is switched on if the input is *NOT* on.

Truth tables can be used to show the behaviour of logic gates. A '0' in the table means off, and a '1' means on.

AND gate:

Input 1	Input 2	Output
0	0	0
0	1	0
1	0	0
1	1	1

OR gate:

Input 1	Input 2	Output
0	0	0
0	1	1
1	0	1
1	1	1

NOT gate:

Input	Output
0	1
1	0

5 The output of the logic gates, or the processor, switches the output device on or off. Common output devices are:

a A **light emitting diode (LED)**, which lights up when a current flows through it.

b A **relay** which is used to switch on a more powerful circuit containing e.g. motors (to provide movement), heaters, lights, locks, or buzzers (to sound an alarm).

Checklist

Are you sure you understand these key terms?

input sensors / processor / output device / light dependent resistor / switch / thermistor / switch / AND gate / OR gate / NOT gate / truth table / LED / relay

Now learn how to use your knowledge

Use your knowledge

1 The diagram shows a simple electronic system:

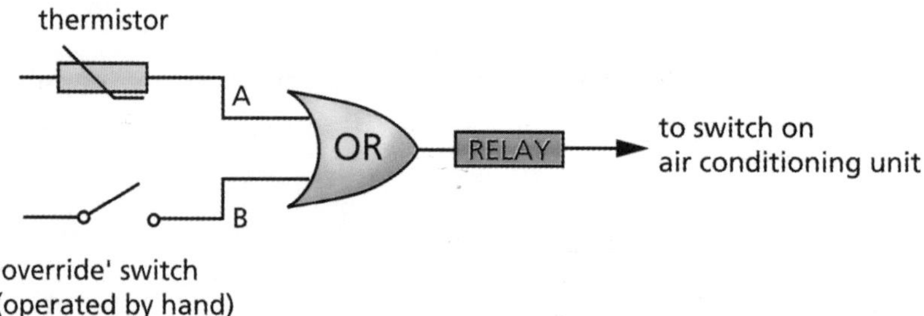

a On the diagram label

1) the input sensors
2) the electronic processor
3) the output device.

b Under what conditions will input A be switched on? Hint 1

....

....

c Under what conditions will input B be switched on?

....

....

d Explain why this system is useful for controlling an air conditioning system. Hint 2

....

....

....

e Why is a *relay* needed? ....

....

f A similar set-up can be used to control a heating system. The heater should be switched on if it is too cold. What change should be made to the system on the previous page for this task? **Hint 3**

....

....

2 The diagram below shows an electronic system used to switch on a security light outside a house.

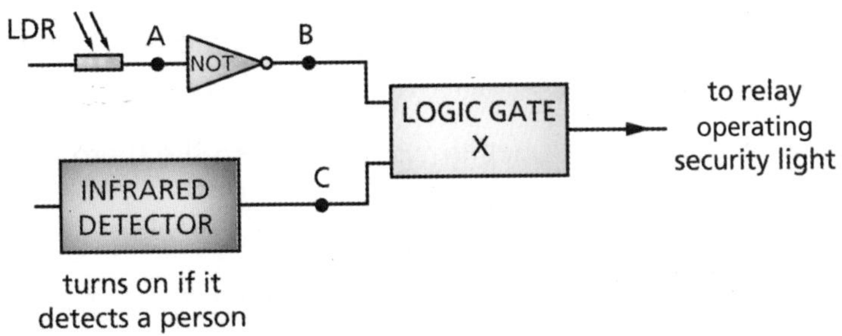

a Fill in the gaps below to say whether the inputs will be **on** or **off**:

If it is dark, input A will be .... , and so input B will be .... **Hint 4**

If a person is detected input C will be ....

b The security light should be switched on if a person is detected when it is dark. What type of logic gate should be used in the box marked 'LOGIC GATE X' to correctly control the security light? **Hint 5**

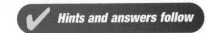

....

Hints and answers follow

Electronic control

Hints

1 Input A is controlled by the thermistor.

Under what conditions will a thermistor turn an input on?

2 The OR gate will turn the output on if either input A or input B is on.

In this case the output will be on if the temperature is hot enough, or the override button is pressed.

3 The air conditioning comes on if it is too hot, but the heater must come on if it is too cold.

So, we need to get an 'on' input from the thermistor when it is *cold*.

What gate can be added after the thermistor to do this?

4 Notice that B is the output of the NOT gate. (The input to the NOT gate is A.)

5 If a person is detected in the dark, inputs B and C will both be on.

What type of logic gate has an output that is on only if both inputs are on?

Answers

1 a) 1) the thermistor and switch 2) the OR gate 3) the relay b) if it is hot enough c) if the switch is turned on d) the OR gate turns the relay on if it is above a certain temperature or an override button is pressed. This is exactly when an air conditioning system should be turned on e) it allows the small electronic current from the control system to switch on a much larger current in the air conditioning unit f) a NOT gate should be added after the thermistor **2** a) A off; B on; C on b) AND gate

Mock exam

1 a

electrical	kinetic	thermal	light
sound	chemical	gravitational	

Choose words from the box to fill in the spaces below:

In a car, ... energy from the petrol is converted into .... (or movement) energy. Some energy is converted into .... energy to power the radio, windscreen wipers, and lights. Some energy is wasted as non-useful energy, mostly in the form of .... and .... energy. **(5)**

b A stationary car puts a force of 12 000 N on the road beneath it.

i) What is the size of the force exerted *by* the road *on* the car?

.... **(1)**

ii) If the area of the tyres in contact with the ground is 0.1 m^2, calculate the pressure exerted by the car on the tarmac under its wheels.

....

.... **(3)**

c Look at the forces acting on the cars below:

3000 N ← A → 1000 N

1000 N ← B → 3000 N

1000 N ← C → 1000 N

i) Which one is moving at constant speed? ....

ii) Which one is braking? .... **(2)**

[11]

 **a** The diagram shows a cross-section through a vacuum flask, used to keep coffee hot for several hours.

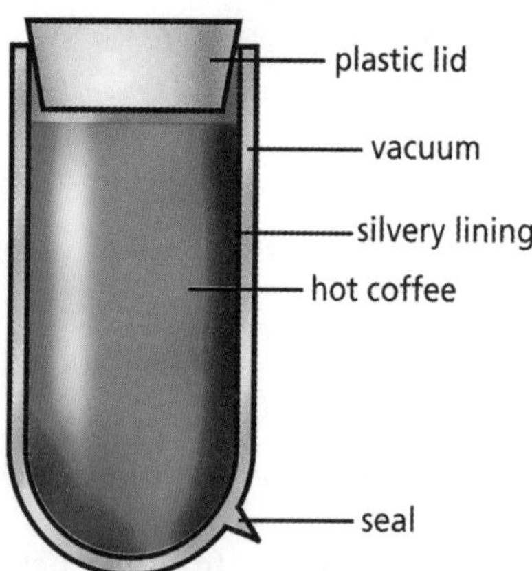

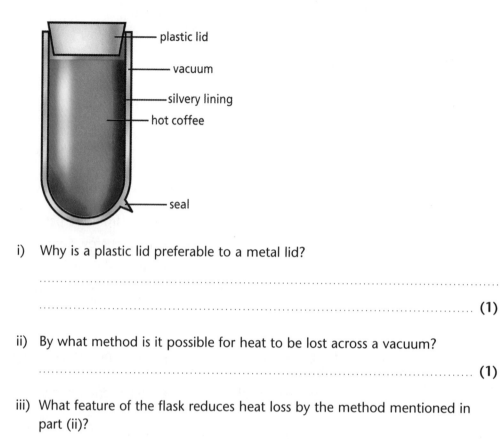

i) Why is a plastic lid preferable to a metal lid?

....

.... **(1)**

ii) By what method is it possible for heat to be lost across a vacuum?

.... **(1)**

iii) What feature of the flask reduces heat loss by the method mentioned in part (ii)?

.... **(1)**

iv) Could the same flask be used to keep an iced drink cool? Explain your answer.

....

.... **(2)**

b Jo and Amber are getting ready to go out on a cold winter's day. Jo says she is going to wear two thin jumpers instead of one thick one, so the air trapped between them will keep her warm. Amber says that air can't be that good at keeping you warm, otherwise you'd be warm enough without any jumpers on, with air all around you!

i) Explain why the air trapped between two jumpers keeps you warm.

....

....

.... (1)

ii) Why doesn't the air around your body keep you warm without needing any jumpers?

....

.... (1)

[7]

3 ***a*** An electric circuit is connected as shown below:

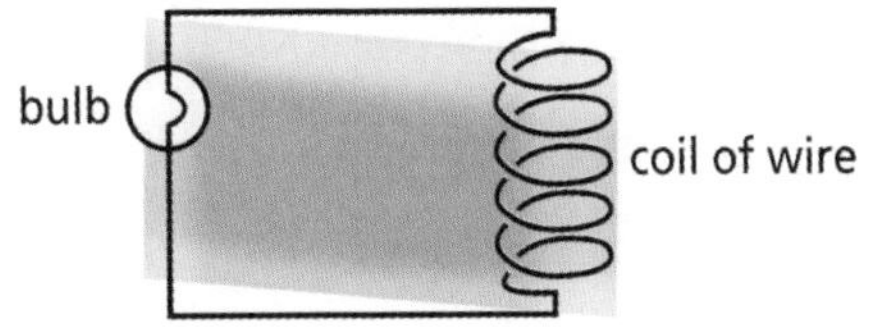

bulb

coil of wire

i) Will the bulb be on or off? .... (1)

ii) Why? .... (1)

b A small permanent magnet is dropped through the coil:

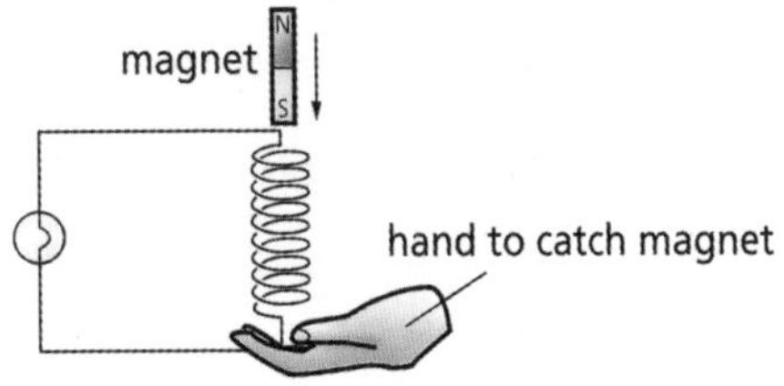

magnet

N

S

hand to catch magnet

While the magnet is moving, the magnetic field felt by the coil is changing all the time.

i) Will the bulb be on or off while the magnet is moving?

.... (1)

Why? ....

.... (2)

ii) Will the bulb be on or off when the magnet stops moving?

.... **(1)**

Why? ....

.... **(1)**

C The diagram below shows a simplified diagram of a gas power station:

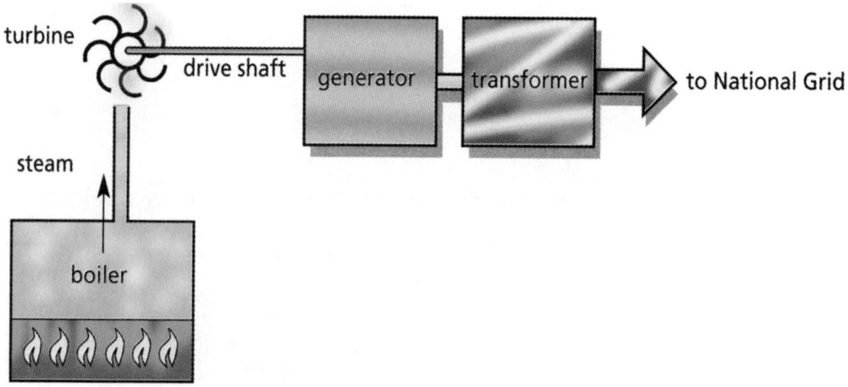

In which part of the power station

i) is chemical energy converted to heat energy?

.... **(2)**

ii) is kinetic energy converted to electrical energy?

.... **(1)**

iii) Inside the generator, the drive shaft rotates a powerful magnet inside a coil of wire.

Why is it important that the magnet is rotating, not stationary?

....

.... **(2)**

iv) What does the transformer do?

....

.... **(2)**

[15]

4 Lucy is going to a friend's house to borrow a book. Her friend's house is about 1000 metres from her home. On the way she stops to buy some sweets. A graph of her movement is shown below:

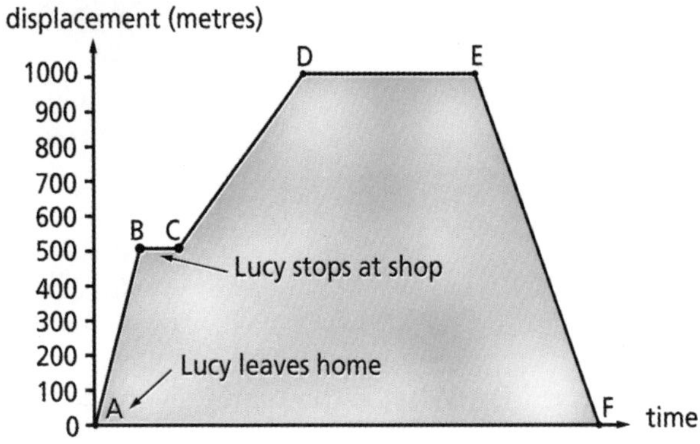

a How far away from her house is the sweet shop?

.... **(1)**

b If it took Lucy 5 minutes to get to the sweet shop, calculate her speed for that part of the journey, in metres per second.

....

....

.... **(3)**

c Does Lucy walk faster before she gets to the shop, or after leaving the shop? Explain how you found this from the graph.

....

.... **(2)**

d Describe Lucy's movement

i) between D and E on the graph

....

ii) between E and F on the graph

.... **(2)**

[8]

5 ***a*** The diagram below shows a circuit used to run the two lights on a bike from one battery:

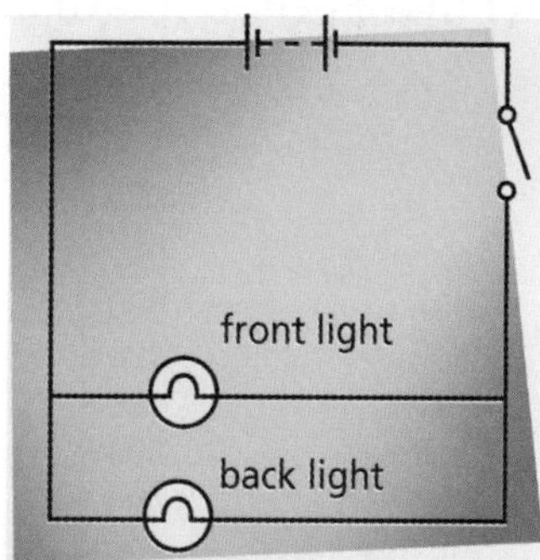

i) What is the name given for this way of connecting two bulbs together?

.... **(1)**

ii) If the back light bulb blows, current can no longer flow through that bulb. Will the front light stay on or turn off?

.... **(1)**

b An alternative way of connecting the bulbs is shown below:

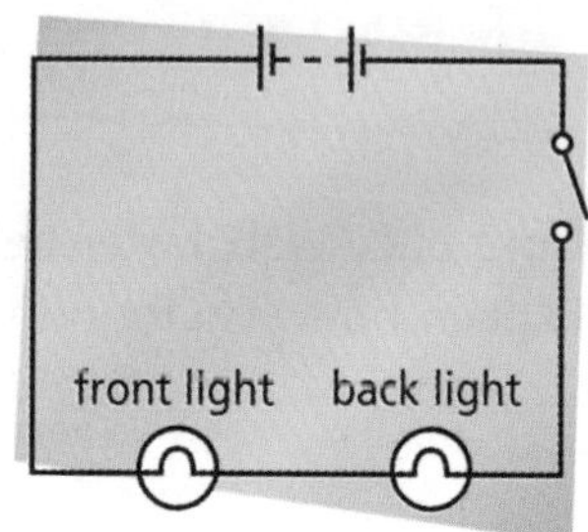

In this case, what happens to the front light bulb when the back bulb blows?

.... **(1)**

C One of the bicycle bulbs is taken into the lab so that its resistance can be investigated.

i) Add an ammeter and a voltmeter to the circuit below to allow the current through the bulb and the voltage across the bulb to be measured. **(2)**

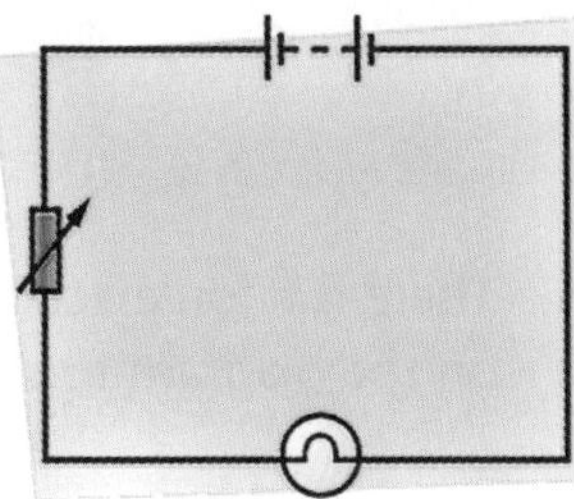

A series of readings are taken of current and voltage, to allow a graph to be plotted.

ii) Why is a variable resistor included in the circuit for this experiment?

....

.... **(2)**

iii) Sketch the shape of graph you would expect to find for the bulb:

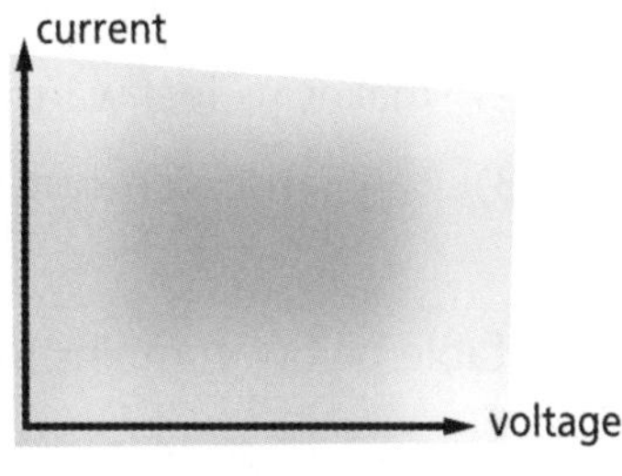

(2)

iv) If the resistance of the bulb remained constant, this graph would be a straight line. Why is a straight line not produced by this experiment?

....

.... **(2)**

[11]

6 ***a*** The diagram shows the Earth and the Sun (not to scale):

i) There is a force acting on the Earth due to the Sun's gravity. Draw an arrow on the diagram to show the direction of this force. **(1)**

ii) If the Sun did not exert this force on the Earth, what difference would this make to the Earth's movement?

....

.... **(2)**

iii) Consider the following four types of waves:

light   infrared   ultraviolet   sound

Which one of these waves

1) cannot travel to us from the Sun? .... **(1)**

2) makes us feel warm when the Sun is out? .... **(1)**

3) can cause skin cancer? .... **(1)**

b The table below contains information about the planets in our solar system:

Planet	Distance from sun *(millions of miles)*	Surface temperature *(°C)*	Surface gravity *(N/kg)*
Mercury	36	350	4
Venus	67	480	9
Earth	93	20	10
Mars	140	-20	4
Jupiter	480	-150	3
Saturn	890	-180	1
Uranus	1800	-210	1
Neptune	2800	-220	1
Pluto	3700	-230	

i) Complete the following statement:

In general, the further a planet is from the sun, the

.... its surface temperature is. **(1)**

ii) According to the information in the table, which planet(s) do not quite fit into this pattern?

.... **(1)**

iii) If men from Earth succeeded in landing on Mars, what difference, if any, would they notice to

1) their weight? ....

....

2) their mass? ....

.... **(2)**

iv) Why is life unlikely to exist on Neptune?

....

.... **(1)**

[11]

Total marks for paper = 63

Answers

1 a) chemical, kinetic, electrical, thermal, sound
b) i) 12 000 N ii) 120 000 N/m^2 c) i) C ii) B

2 a) i) plastic is a poorer conductor of heat ii) radiation iii) the silvery lining iv) yes, because it will prevent too much heat getting in to warm up the drink
b) i) air is a poor conductor, and trapped air cannot transfer heat by convection either ii) if the air is not trapped, you will cool down by convection, or draughts

3 a) i) off ii) because there is no battery or other power supply
b) i) on, because the changing magnetic field induces a voltage in the coil, which makes a current flow in the circuit ii) off, because the magnetic field is no longer changing, so there is no induced voltage
c) i) boiler ii) generator iii) rotating the magnet gives a changing magnetic field, which induces a voltage iv) increases the voltage of the electricity leaving the power station

4 a) 500m b) 1.7m/s
c) before she gets to the shop, because the gradient of the graph is steeper for that part of the journey
d) i) she has stopped (at her friend's house) ii) she is returning home (back to where she started)

5 a) i) in parallel ii) it stays on b) it goes out
c) i) see below ii) so that the current and voltage values for the bulb can be varied iii) see below iv) as the bulb gets hotter its resistance increases

6 a) i) see below ii) the Earth would not orbit the Sun, but would continue in a straight line into space iii) 1) sound 2) infrared 3) ultraviolet
b) i) lower (or colder) ii) Venus (and/or Mercury) iii) 1) weight would be less than on Earth (as the gravity on the surface of Mars is less) 2) mass would be the same as on Earth iv) due to the extremely low temperature

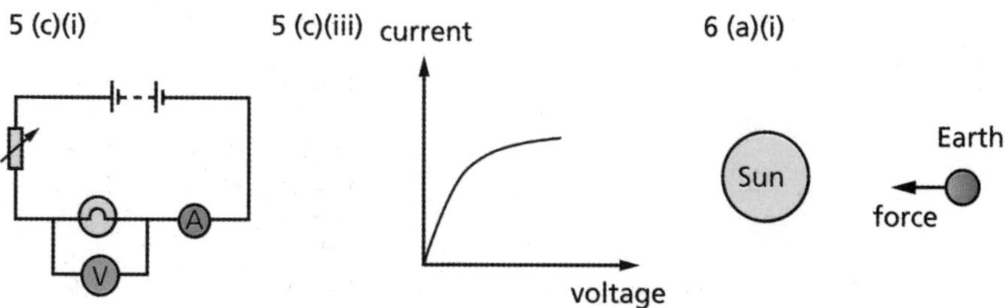